A UNIVERSAL MODEL

FOR

D1744486

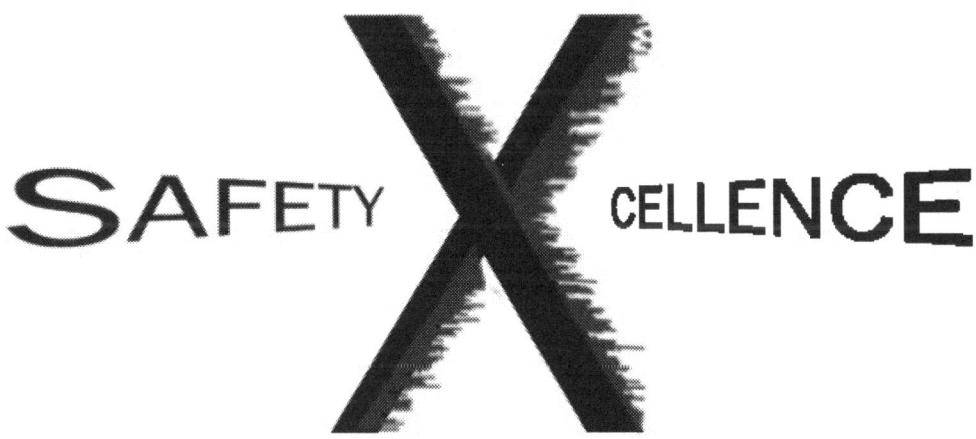

SAFETY **X** CELLENCE

"Excellent organizations frequently achieve exceptional safety results in the absence of any 'visible' safety program."

"Excellent safety performance cannot be attained in a generally poor organization."

— D. A. Weaver, Inductee
International Safety Hall of Fame

LARRY L. HANSEN, CSP. ARM & DANIEL F. ZAHLIS

ISBN 10: 1-933817-11-9
ISBN 13: 978-1-933817-11-8

First Printing July 2006

Published in the USA by
Profits Publishing of Sarasota, Florida

SPEAKING OF SAFETY, INC.
Baldwinsville, New York

For Janice Ann, who waited for me her entire life…
I so hope she's not disappointed.

ABOUT THE AUTHORS

 Larry L. Hansen is Principal of "L2H Speaking of Safety, Inc.", a Safety Excellence consultancy, and Vice President of Active Agenda, Inc. Larry holds a BS degree in Operations Management from the University of New Haven, an Associate in Risk Management Designation from the Insurance Institute of America, and is a Certified Safety Professional.

Larry has over 36 years of professional experience in Safety Management, Organizational Performance, and Strategic Consulting. He is recipient of the Irwin/Proctor & Gamble and ASSE/Veterans of Safety Professional Best Paper awards, for his work entitled: Safety Management: A Call for Revolution."

 Larry has authored over 35 articles on safety management, fourteen of which have been selected as 'cover/features' for national magazines including: "Re-braining Corporate Health and Safety;" "Twelve 'Unlegal' Ways to Slash Workers' compensaiton Costs;" "What's Your Organization's Loss Management IQ? Survival Skills for the Safety Professional;" "Stepping Up to Operational Safety Excellence" and "The Architecture of Safety Excellence." He is author of the book: "ROC Your Organization: Fifty-two Ways to Instigate Radical Organizational Change for Safety Excellence" containing change initiatives designed to help organizations achieve peak safety performance. Larry resides in Baldwinsville, New York and can be reached (when not shoveling) at (315) -383-3801, via e-mail at LLHSOS@dreamscape.com and on the web at: www.L2Hsos.com

 Dan Zahlis is Founder and President of Active Agenda, Inc., a Risk Management software company. Dan is a graduate of California State University Fresno with a BS degree in Health Science, Occupational Safety and Health. He went on to attend McGeorge School of Law where he left before it was too late!

Dan began his career in occupational safety with Chemical Waste Management where he worked as a safety engineer and eventually assumed the responsibility for the safety and health of the nation's second largest hazardous waste TSD facility. He went on to work with Dole Food Company, and the Häagen-Dazs Company where he gained valuable insights regarding organizational performance, strategic change, and the importance of measurement systems, which encourage frequency (reporting) as a means of decreasing severity (cost) . Dan authored, and received the American Society of Safety Engineers' - Veteran's of Safety, Professional Paper award for his work, CAUTION, Beware of OSHA Statistics, and his follow-up article: The Hidden Agenda, which was selected as 'Cover/Feature' of Professional Safety magazine. In 1995, Dan began focusing his efforts to bring Active Agenda® an automated risk management operating system to the marketplace. Active Agenda® leverages open source concepts, and today's technologies to help companies mine upstream data ('CAUTION') and streamline the process of risk administration through collaboration and integration ('The Hidden Agenda') .

Dan can be reached at (559) 916-6106 or by email at dzahlis@activeagenda.net.

THE THINKER

Reprint from Industrial Safety & Hygiene News -September 1993

He says safety pros are in a comfort zone.

Larry Hansen, a Safety Excellence and Organizational Performance consultant, is one of the people stretching the envelope of safety management thought. He answered questions put to him by Industrial Safety & Hygiene News.

What's the basis for the article you wrote titled, "A Call for (R) evolution, " which appeared in the March "Professional Safety Journal? "

In June of 1993 I celebrated my 25th anniversary in the loss control profession. As I pondered my career, I realized that not all that much has changed. The battles are different but there's not much progress. The realities are:

1) The long-term trend in workplace injury rates hasn't improved substantially over the last 20 years, and some of the studies suggest that the numbers we're looking at are really worse than we suspect due to under reporting.

2) NIOSH studies in the' 70s and' 80s comparing companies with good loss experience versus those with poor loss experience found" no significant difference" in accident results based on" traditional" safety programs.

These studies found that companies relying heavily on safety directors and safety committees actually produce worse results. Companies that rely more on staff administration end up usurping line management responsibility. Those who can control accident exposures abdicate responsibility to those who can't(the safety director) .

As consultant Paul Coderre says," Safety and business have traditionally been mutually exclusive. Continuing this separation in today's economic climate will be fatal."

My desire is to make a" positive difference" by bringing some of these critical issues to light and causing American industry to re-evaluate its approaches and values.

What's the most important message you're trying to get across?

The message is easy. Getting organizations to accept it is the hard part. To paraphrase President Clinton," It's MANAGEMENT stupid! "

One reason industrial safety has had limited success is the perception that safety is" technical" - not a managerial/ organizational issue. But the bottom line is this: Like quality defects, excess costs, and high scrap, employee injuries(human scrap) are just another outcome of poorly designed management processes.

Management types don't generally like to hear this message, but I believe it's the truth. Those who eventually accept it and make serious efforts to change their organizational philosophies, methodologies, and processes reap huge rewards. In well-managed companies, safety isn't a program, it's how things are done routinely and continuously.

You say safety leaders are needed to help bring about this change. Where are they?

The vast majority of safety practitioners in industry today can be put in one of two categories: 1) Those who haven't got a clue! and 2) Those who haven't got a chance!

This profession has been populated many years by folks who have been schooled with a real focus on technical safety. At this point they are comfort-able, unwilling to take risks, suspect of new ideas, and not learning.

The second group, those who haven't got a chance, are actually enlightened individuals. They're newer to the business and recognize the need for change. Unfortunately, these people are trapped in organizations that don't recognize or believe safety is a management function, not simply a matter of compliance or technical efforts.

There is a third category. Those safety practitioners who are succeeding by applying organizational and management approaches to their safety challenge. Unfortunately, these are a minority.

Your recommendations for improvement include employee empowerment. Specifically what type of empowerment is needed?

It's important to keep in mind that before an organization can pursue "empowerment" it must first create "enablement."Employees must be knowledgeable and capable of doing the job. Unfortunately, United States industry does not effectively develop an employee's ability to think.

Once employees are enabled, they can then be empowered. That is, provided with authority to take independent actions in problem solving and process improvement. This requires "trust" on the part of management-something that is not frequently given in organizations. For empowerment to work, values have to replace rules. Rewards for risk-taking have to replace fears of failure.

One of the most successful ways to fuel empowerment is through gain sharing-linking financial rewards to successful efforts. Unfortunately, my experience finds that reward systems are really screwed up in most workplaces. Far too often, only executives reap the rewards (bonuses and perks) for innovations and productivity increases driven by the first line.

I also see where employees suffer and/or are called upon to pay the price (called "right sizing, down sizing, out-placing") caused by short-sided "QTQ" (quarter-to-quarter) management practices that wreck organizations.

I find that employee relations policies today put too much emphasis on "measuring" people rather than "valuing" them. These policies are reflected in a number of our outcomes including our injury and illness rates.

What's your basis for saying "employees fail due to management? "

My reason for believing this is based on the works of W. Edwards Deming and Tom Peters.

I happen to believe in Deming's common cause and special cause theory which equates 90 percent of our system outcomes (including accidents) to common causes, those causes embedded in the systems designed by management. Only 10 percent of outcomes are due to special causes of individual behaviors.

Tom Peters, in turn, speaks of "management enlightenment" as "a blinding flash of the obvious."In safety, this phenomenon exists in the linkage of accident causes with their true organizational sources. Most executives would readily agree that "management" has the responsibility for: design of products, selection of employees, specification of materials, development of policies, drafting of procedures, specifying machinery, planning work schedules, controlling work environment, and shaping organizational culture by all of these.

But when accidents occur from exactly these same factors, "the blinding flash of the obvious" is: "Damn those careless employees that keep having accidents! "I don't buy that. I believe that causes of accidents in the workplace are embedded in the management process, not in employee behaviors. It's a "control issue" and management controls all the loss-producing sources.

As Pierre Morrisseau says, "A company is a system which can be broken into three components-physical (machinery), human (employees), and social organization.

CONTENTS

A UNIVERSAL MODEL FOR SAFETY X-CELLENCE.

PROLOGUE

As I approached the mid-point of my career I was confronted by two significant realizations. The first was that the traditional '3E' safety strategies (Enforcement, Education, and Engineering) that I had been trained in, and was employing with my clients…just weren't working. The second was that after 20+ years in safety management, I had not made a single meaningful contribution to the profession. On June 30, 1991 I made two decisions to remedy both those situations.

The first was a self directed career change from Safety Consultant to 'Organizational Performance Consultant', and since I was a straight 'C' student in High School English, scored in the bottom quartile of my SATs, and flunked Freshman remedial English in college (twice), the second was to write a professional article. The following day, I put pencil to paper…and 18 months later I finished a manuscript entitled: "Safety Management: A Call for Revolution."

In 1993, Professional Safety magazine selected and published that work as the cover of its June issue…and the 'Revolution' was underway! Ultimately that article received 'Best Paper' awards from ASSE, The Veterans of Safety, and The Minerva Educational Institute. What I set out to accomplish as an end goal, hence became the start of an ongoing effort to challenge tradition, create change, and bring value to the Safety and Health profession.

Two years later, another 'challenger' of the status quo named Dan Zahlis received the same ASSE Best Paper award for his article: "CAUTION: Beware OSHA Statistics", and followed that with another 'Professional Safety' feature entitled: "The Hidden Agenda". These articles addressed the need for, and value of pursuing collaboration, and employing technology to achieve safety excellence. I clipped both for my 'TGRHHST' file (this guy really has his s---- together).

In 1998, I received a telephone call from Don Eckenfelder, a friend and a respected safety management consultant. Don had been engaged by an investor group to do a 'due diligence' assessment of an emerging risk management technology company in Clovis, California. After meeting the founder, and completing that assessment, Don called me and said: "Larry, I've just met the smartest safety guy on this planet…a guy named Dan Zahlis". I immediately went to my 'special file' pulled the articles, and said: "Do you mean this Dan

Zahlis?" Don went on to tell me that in his discussions, he mentioned my name, and upon hearing that, Zahlis said 'wait a minute', went to his filing cabinet and pulled a file containing my 'Call for Revolution' article, and said:" Do you mean this Larry Hansen?"

We had discovered and tracked each other's work for a number of years...yet had never formally met. That chance happening led to Dan and I meeting a few weeks later, where we confirmed our common beliefs and values, and agreed to collaborate in bringing positive change to the safety profession. The vehicle we would employ was 'ACTIVE AGENDA' (see Section 14), a risk management technology capable of facilitating culture change that Dan had been evolving for a number of years.

Since that meeting, Dan and I have collaborated in the evolution of Active Agenda, and have authored over 35 feature articles on safety management, excellence strategy, and risk management technology, including 14 cover articles for national magazines such as; Professional Safety, Occupational Hazards, Occupational Health & Safety, and Risk Management.

The writings assembled in this book have one key objective, and that is to provide insight on the core question of safety excellence: "Why are some companies able to achieve peak safety performance, while so many others struggle to maintain mediocrity?"

Excellence companies, (the best) aren't just luckier than the rest; they differ from their lesser performing peers, in two very distinct ways. First, they achieve significantly different (better) results, and second, they do significantly different (best practice) things...and, most importantly, they realize the two are linked! These companies capitalize on what organizational performance researchers from the Gallup Organization have recently confirmed: "Excellence isn't the opposite of mediocrity; excellence is different!" High performing organizations 'THINK' differently about safety...they have a 'Safety Excellence' mindset!

This series of essays and published articles explore this 'Safety Excellence Mindset'...those critical differences in how leaders in safety excellence companies think...and as a consequence what they 'DO', to achieve world-class results.

The book begins with some 'baseline' activities (Mind-Shifting exercise, and Brain-Drain quiz), intended to surface current beliefs, and expose prevailing 'Wiz-dumbs' in safety (wrong-headed thinking that impedes progress in the right direction) that may be impacting current results in an organization.

Next, it presents a 'Universal Model of Safety Excellence'; composed of peak performance elements drawn from a broad perspective of operational excellence literature. This model identifies seven core elements of a Safety Excellence Culture, i.e., the 'change targets' requisite to achieving peak safety performance.

The ensuing chapters (articles) sequentially identify and align these elements into the 'Universal Model' of safety excellence, exploring the critical importance of these elements in successfully achieving World-Class safety.

We do hope you enjoy these 'reads and rants', and most importantly, find them valuable in your QUEST for Safety Excellence.

A UNIVERSAL MODEL FOR SAFETY X-CELLENCE.

SECTION 1

Organizational Change

"A person who cannot change his mind, cannot change ANYTHING."

~ George Bernard Shaw

"Consider how hard it is to change yourself, and you'll understand the chances you have of trying to change others."

~ Jacob M. Braude

"Change Daily"

~ Joe Boxer, Inc.

REQUISITE TO EXCELLENCE

"We don't need change agents, we need change insurgents."

~Robert Reich
Former Secretary of Labor

Brian Tracy, author of "The Creative Manager" says that between you and your ultimate goal lies a rock—a significant impediment that prevents you from successfully achieving your objective. In my work with companies striving to become World-Class safety organizations, I've learned that the greatest impediment to achieving safety excellence is an inability to create real sustainable change in their organizations. By sustainable, I don't mean the ordinary run of the mill type of change (program today; gone tomorrow), but rather, change of the frame bending, mind altering, type, I call the 'ROC'… Radical Organizational Change.

As you pursue transformation in your safety process, remember that one very powerful force is actively working against you --your current mindset! Will Rogers called it the greatest impediment to success: "What you know that just ain't so". For real change to manifest in an organization, you need open your mind to new possibilities, remove the filters of convention, and resist the temptations of silver bullet solutions, i.e., what's fast, easy…and so often wrong!

Unfortunately, much of the conventional thinking known as the 'Safety Program' is comprised of ideas that run contrary to sound management principles, and/or have lost relevance over time. When results have stagnated or are trending in the wrong direction (as incident rates have in many organizations for years), the last thing you want to do is become more efficient! A strategy that calls for more of the same--only faster, quicker, harder and cheaper isn't the answer!

After an exhaustive study of today's high performance companies, Marcus Buckingham, lead consultant for the Gallup organization and co-author of " First Break all the Rules" concludes: "Excellence is not the opposite of mediocrity…Excellence is different! "

In business and safety, there are no quick fixes, but when you commit to doing the right things, you discover

2

that there can be rapid returns! In a Fast Company magazine interview, former Secretary of Labor Robert Reich, said: "We can no longer just be change agents in our organizations, we must become change insurgents."

James Champy, author of, "Re-engineering Management", says that today's leaders must have: "A mind perpetually ready to revolt against its own conclusions." To succeed, you must dare to be different; dare to question the status quo; dare to "ROC Your Organization" into safety excellence!

MIND-SHIFTING TO EXCELLENCE

In my safety excellence seminars, I frequently engage management groups in a short thinking exercise (ouch) to expose conventional paradigms, and determine if they could value from a 'mind-shift', i. e., a change in perspective on the 'purpose, mission, and value' of safety in their organization. It hence seems fitting that you, as a reader, be offered the same opportunity. Complete the following exercise to determine if you personally could value from a 'mind-shift' to Safety Excellence.

EXERCISE

Instructions: Listen to the little voice in your head, and carefully follow these instructions (use borders to jot calculations if necessary, and record your answers below) :

1. Select a number from 1 to 10.

2. Multiply that number by 9.

3. If the resulting number is a 'two digit' number, add them together.

4. Now, subtract 5.

5. Determine which letter of the alphabet corresponds to this number: (Ex. 1=A, 2 =B, 3 = C, 4 = D, etc.) .

6. Think of a COUNTRY that starts with that letter. (Note below)

7. Remember the last letter of that country.

8. Think of an ANIMAL that starts with that letter. (Note below)

9. Remember the last letter of that animal.

10. Think of a FRUIT that starts with that letter. (Note below)

COUNTRY_____

ANIMAL_____

FRUIT_____

Now, proceed to the next page and answer the following three questions:

QUESTIONS TO PONDER

QUESTION 1:

"Did your little voice say: 'This is stupid…What does this have to do with safety excellence? Why am I doing this dumb exercise?'"

QUESTION 2:

"Did you tell your little voice to: 'Shut Up'…and continue anyway?

QUESTION 3:

"Do you do safety the same way, and for the same reasons?

"Much of American business doesn't seem willing or equipped to address directly what is often the real core of operational problems --MINDSET."

~ James Champy, Author
Re-Engineering Management

Now, proceed to the next page and answer these truly relevant questions about 'Safety Excellence'.

BRAIN-DRAIN

IMPORTANT: Read instructions carefully before answering these questions.

Instructions: This quiz can be taken in one of two ways: you can test your own beliefs by answering personally, or you can capture an organizational perception by answering the following questions as you believe the vast majority of line managers and supervisors in your organization would answer them. Personally, I think the organizational perspective will provide the greatest value…. as ultimately that is what need change for an organization to succeed. – Your choice!

1. **True or False** - The primary objective of a safety program is to prevent accidents.

2. **True or False** – Accidents are unplanned, unexpected, unintended events, and as such, are highly unpredictable and very fortuitous occurrences.

3. **True or False** – Three key differentiating features of an effective safety program are 1) Size of the safety budget; 2) A safety committee; 3) The quality & quantity of safety rules.

4. **True or False** - An effective safety program targets 'Efficiency'…Getting people to 'Do Things Right'.

5. **True or False** - The number one (#1) cause of accidents in organizations is unsafe employee acts.

6. **True or False** – To achieve a high level of safe work behavior, an organization must have comprehensive policies, safety rules, performance standards, and safe operating procedures.

7. **True or False** – A uniform and equitably administered progressive disciplinary policy and procedure is critical to raising the level of safe work behavior in an organization.

8. **True or False** - To elevate the level of safe practice in an organization, a company must place an increased emphasis on employee safety training.

9. **True or False** – Over the long term, an organization that maintains its OSHA recordable incident rates to be equal to or slightly better than industry average, will continuously reduce its loss costs.

10. **True or False** - Zero Accidents is the ultimate goal of safety excellence.

BRAIN-DRAIN

Answers

1. **False** - The primary objective of Safety Excellence is Process Improvement. The quality field taught us in the late 70's that it is ineffective and unproductive to try and 'inspect out' or 'observe/discipline out' defects at the end of a process line. The safety profession needs to learn from this, and re-focus efforts from 'inspecting out' hazards and at risk behaviors on the shop floor to 'designing in' safety in the executive boardroom and on the engineer's drafting tables.

2. **False** – The types, rates, trends, and systems that generate accidents are highly predictable…the only unknowns are: ' Who and When'! Our national BLS incident rate trends, and the cost and occurrence data evident in most organization's OSHA logs and insurance loss runs and premium bills make this amply clear.

3. **False** –Studies by both public and private organizations, including large insurers, consulting firms, the Department of Energy, and NIOSH studies of safety program effectiveness confirm that: 1) Size of Safety Budget; 2) Existence of a safety director/committee, and; 3) Quality and Quantity of Safety Rules, are issues having minimal impact on safety results. These are not the differentiating features of high performance safety organizations.

4. **False** - An effective safety process targets 'Effectiveness'…'Doing the Right Things, rather than 'Efficiency'…Doing Things Right. Excellence is not the opposite of mediocrity…Excellence is Different. Excellence companies go beyond the traditional '3E' program strategies of, Education, Enforcement and Engineering—they focus on Culture (values), Leadership (practices), and Organization (structure) which dictate Organizational Behaviors…the strategies of excellence.

5. **False** - The number one (#1) cause of workplace accidents is the reason(s) behind unsafe employee behavior, a/k/a 'the good reasons for poor performance': executive values, management practices and process design.

6. **False** – To achieve a high level of safe behavior, an organization must emphasize Consequence Delivery by line managers. Safety programs (the policies, rules, and performance standards) developed and administered by the staff (safety/HR) organization, are antecedents, which by definition have limited impact on employee work practices.

7. **False** – Disciplinary policies and procedures can be effective in stopping unsafe work practices, but performance research on human behavior is clear, punishment does not 'increase' the level of safe

behavior in an organization. Reinforcement, predictably and continuously delivered by line managers, is required to increase the level of safe/desired behavior in an organization.

8. **False** – Numerous studies, including those conducted by the US Postal Service, a national insurer, and the NHTSA have confirmed that operator training (a behavioral antecedent) does not have a significant impact on safe behavior. Most accidents are the result of performance deficiencies (not doing) …rather than knowledge deficits (not knowing). Employees generally understand 'safe vs unsafe', i.e., the rules. When employees engage in at-risk behavior, it is usually due to a stronger set of 'unwritten rules' at work in the organization…its culture!

9. **False** –Loss trends over the past 30+ years (and continuing) have clearly established that flat incident rates generate significant increases in loss costs due to loss development, the expansion of covered perils in workers' compensation, and the aggressive pace of medical inflation. Injury management strategies can impact the low hanging fruit of cost containment, but to significantly impact loss costs over time, the rate and severity of accidents must be addressed.

10. **False** - According to Paul O'Neill, past CEO of Alcoa: "The absence of accidents in no way confirms the presence of safe." 'Zero Accidents' (recorded) is often the result of luck and manipulation (a/k/a creative classification and record keeping) especially when these numbers are the basis for manager incentives and bonuses). The real 'measurable goal' of safety excellence is visible and quantifiable behavior (of executives, managers and employees) —best expressed by Rob Ryan's definition:*" Safe is: 100% of the people, working 100% safe, 100% of the time!"*

SCORING YOUR BRAIN-DRAIN QUIZ:

To calculate your Safety Excellence Mindset score, and determine what it means, multiply the number of FALSE answers you selected on the Excellence Mindset quiz by ten (10) points and total your score. Find your rating below.

IF YOUR SCORE IS (0, 10, 20, OR 30) :

Your organization is deeply mired in the myth-conceptions of traditional safety, and these significantly impede your current performance. Safety efforts are highly reactive and mostly focused on patching conditions (compliance) and fixing employees (attitudes), neither of which are sustainable safety excellence strategies.

IF YOUR SCORE IS: (40, 50, OR 60) :

Your organization is starting to emerge, but is still predominantly locked into a traditional safety mindset. There are some champions for change and pockets of progressive thinking starting to stir and have localized impact in your organization. Safety is perceived as a program, and is generally something managers do when scheduled, assuming of course, that they have nothing more important to do at that moment. Accidents are mostly the result of employee carelessness; hence initial training, retraining, remedial training, and ultimately discipline are the predominant strategies employed. You often feel like you're on a treadmill, working hard but making little real progress. Your results at best, are 'flat-lined'…and at worst, fluctuate widely year to year.

IF YOUR SCORE IS (70 OR 80) :

Your organization has been enlightened, and has discovered the value of progressive safety management strategy. You are well along on your journey to excellence, and due to strong values and bold leadership are improving day by day. You recognize that management systems harbor most root causes of accidents, and hence focus on organizational change and process improvement as the proper targets of safety excellence. You are committed to driving change in the real elements of safety excellence, value systems and leadership practices. Your destination is near—don't give up now!

IF YOUR SCORE IS (90 OR 100) :

You've arrived! You are a safety excellence organization! You have an excellence mindset which places you in the top ten percent, a/k/a a world class organization. Safety in your organization is not a program, but rather your process…how you do business. Safe is naturally and seamlessly integrated into all key business systems, processes, and management decisions. Congratulations on fighting the good fight, and staying the course of safety excellence. You are one of the elite! Stand tall and be proud!

SAFETY: BETWEEN A 'ROC' AND A HARD PLACE

LARRY L HANSEN

When called upon to facilitate strategic planning in organizations seeking to attain safety excellence, I often begin the first management session by asking the 'two core questions' of excellence…(many sessions, unfortunately end here as well) . The first question is: "How many here today want 'to be' a safety excellence organization?" In response, most arms come flying out of their sockets! I tactfully then present the second question: "How many here today are willing 'to do' safety excellence?" Suddenly, 90% or more of the audience remember their very first golf lesson, "head down, elbow glued to the hip!"

I then pass on two insights on organizational change from Peter Drucker: "The only things that evolve in an organization are disorder, friction, and malperformance," followed by his second insight: "Everything degenerates into work." The bottom line is, change (for the better) doesn't just happen…change is work   (W − f X d) !

Brian Tracy, author of "The Creative Manager" says that between you and your ultimate goal lies a rock, a significant impediment that prevents you from achieving your objective. To be successful, you must find a way over, under, around or through your rock…or as an option, blow it up!

In my work with companies striving to become safety excellence organizations, I've learned that the greatest impediment to achieving safety excellence is an inability (or unwillingness) to create sustainable organizational change. By sustainable, I don't mean the ordinary run of the mill type of change (a little off the top), but rather, change of the frame bending type I call a 'ROC' - (Radical Organizational Change). To truly impact organizational performance, we need change 'what's inside the boxes', the basic beliefs, prevailing assumptions, and core values of the organization…its culture. James Champy, author of 'Re-engineering Management' observes: "Much of American management doesn't seem willing or equipped to address directly what is often the real core of operational problems…MINDSET!" And, Jack Stack, author of 'The Great Game of Business' cautions: "The practice of management I've discovered is filled with myths that are guaranteed to screw up any factory or company. The real secret of an effective manager is to learn how to ignore them."

TRADITIONAL 'WIZ-DUMB'

Both these individuals identify what I've come to call the conventional 'Wiz-dumbs' of business, (Wrong headed thinking that impedes progress in the right direction) . In safety, traditional Wiz-dumb proliferates, has gone unchallenged, and continues to impede efforts and results. Some of the more prominent 'myth-conceptions' and 'Wiz-dumbs' inhibiting performance follow.

TRADITIONAL 'WIZ-DUMBS'

Note: As you read this list, think of the predominant manager mindset in your organization. How would your managers respond to these statements…True or False? See answers at end of article.

- The primary purpose of a safety program is to prevent injuries.

- Accidents are caused by unsafe acts.

- Zero accidents reported equals a SAFE process.

- Accidents are unplanned, unforeseen, fortuitous events.

- Safety is the responsibility of the Safety Function.

- A focus on regulatory compliance will minimize loss cost.

- Repetitive at-risk behaviors indicate a training deficiency.

- Disciplinary actions are necessary to increase safe work practices.

If managers in your company believe each of these to be 'TRUE', you have an immediate need to "ROC Your Organization." These counter-productive beliefs and performance impeding mind-sets must be challenged… and changed, if anything other than mediocrity is your desired objective!

LET'S ROC!

Here are six 'good reasons' for challenging 'poor performance' in your organization:

1. First, because safety makes solid 'financial sense'. According to Peter Drucker: "The first duty of a business is to survive, and the guiding principle of business economics is not the maximization of profit, it is the avoidance of loss." A manager's primary duty is NOT production (output) …it is 'productivity'! (Managing both inputs and outputs) . Managers must focus on the middle lines (costs, expenses, and L. O. S. S.-- Lack of Safety Strategy), because that's how they create the bottom line (margin) . You measure below the line, but you manage above the line. It's all about minimizing loss, by safeguarding and optimizing Human Capital.

2. Second, because good business and corporate citizenship dictates that it's the right thing to do! I personally like the way Charles Jones of EDS puts it…kinda hard to argue with his logic! *" If you don't demand something out of the ordinary, you won't get anything but ordinary results."*

3. Third, because the numbers aren't good. In spite of incident rates getting lower, (no where near excellent) the more important numbers (those with 'dollar signs') are degenerating rapidly!

4. Fourth, because what we're doing isn't effective. A media poll of safety practitioners asking about the effectiveness of current safety efforts revealed:

Safety Programs
(Percent Who Agree)

➲ Are Reactive	83%
➲ Fire Fighting	83%
➲ Quick Fixes	79%
➲ Lip Service	79%
➲ Committees Ineffective	74%
➲ Are Isolated	70%

- ISHN White Paper Poll

5. Fifth, because National Safety Council researchers (Planek and Fearn, Professional Safety, October 1993) examining and ranking effectiveness of safety activities identified the following as 'the bottom' of the list…it has an uncanny resemblance to what many organizations put on the top of theirs!

Least Effective Practices

1. **Employee Contests**
2. **Safety Records**
3. **Injury Records**
4. **Enforcing Resolutions**
5. **Incentives**
6. **Off-The-Job Programs**
7. **On-Site Health Care**
8. **Comparing Records**
9. **Investigating Accidents**
10. **Complying with Standards**

6. And, finally, because Human Resource executives of the Corporate Leadership Council - Advisory Board, when asked what activities could be out-sourced, responded: "Safety—no perceived value added." ("Hey, that's my Boss!")

Former Secretary of Labor, Robert Reich in Fast Company magazine contends that: "Success doesn't require change agents; success demands change insurgents." Organizations can only change the safety results they achieve, by changing 'what and how' they do safety.

CHANGE STRATEGIES

The Success Equation = CEO…'Change Everything Often! '

In an organizational safety context, there are three types/levels of change, each of which addresses a different target, and correspondingly, has an increasingly greater impact on performance and results. These are:

- **Level I: Transitional (minor) Change:** Initiatives which focus on changing working conditions and behaviors, a/k/a- Safety Programs.

- **Level II: Transactional (moderate) Change:** Initiatives which focus on changing organization (structure and roles), process (systems), and management (practices) …a/k/a – Safety Management and,

- **Level III: Transformational (major) Change:** Initiatives which focus on changing organizational culture (values) and executive actions, a/k/a - Safety Leadership.

To 'ROC' an organization to excellence, leaders must target Level II and Level III change; as these harbor the headwater causes of loss in an organization. Losses are driven by…and consequently best controlled by addressing the five advanced strategies of safety: Culture (values), Leadership (actions), Organization (structure), Process (systems) and Management (behaviors) .

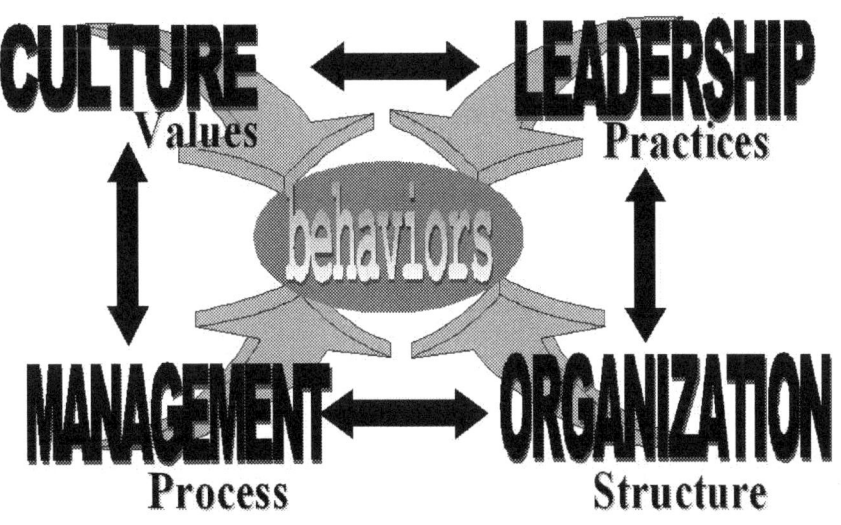

THE 'ROC' TESTS

As the hardness of a gemstone can be tested on a MOHS (1 to 10) scale, so to can the potential impact of 'ROCs' on an organization using a (1 to 7) impact scale. Initiatives designed to instigate organizational change in performance and results can be gauged against these seven critical characteristics:

A 'ROC' is: An organizational safety improvement initiative that:

1. Targets change in one of the five advanced strategies of safety excellence: (Culture; Leadership; Organization; Process; and Management Behavior.

2. Is 'non-traditional' in concept…not 'S. O. S. S.' (Same old Safety S __!) …Stuff.

3. Emotionally engages, and unifies the organization—builds collaboration.

4. Challenges the status quo - recognizes that current success is an impediment to future success.

5. Forces the 'real issues' onto the table…the sacred cows and 500 lb. gorillas!

6. At first blush, appears a little 'off the wall'…but, upon closer examination, is!

7. Passes the organizational 'Eye Test' i.e., When proposed it evokes these visual reactions: (look into their eyes and discover the truth!)

    -   Employees: 'Beam' – because, they see it's right!

    -   Managers: 'Wince' – because, they see it's work (for them) .

    -   Executives: 'Furl' – because, they see themselves in the 'crosshairs'.

Note—If you ever develop a ROC that scores 'seven' on this scale you most likely are on to something that will 'Nuke' the organization!

NO QUICK FIXES

Although there is no such thing as a 'quick fix' in business…or safety, there is the very real potential for 'rapid returns' when the same amount of time, money and resources are shifted from traditional (low impact) strategies to the high impact advanced strategies of excellence. "ROC Your Organization" is all about refocusing an organization from 'efficiency'—doing things right, i.e., (safety programs) to 'effectiveness' – doing the right things (safety management and safety leadership) . It is a conceptual (and practical) approach to help safety champions and their leaders, transform 'how' safety is done in an organization by promoting new thinking, encouraging different actions, and enabling better outcomes.

Practitioners who adopt a 'ROC' mindset, become change insurgents, and take those overt and covert actions requisite to creating a Safety Excellence organization. In the words of Secretary Reich: "Change doesn't happen top down; change happens from where ever you are." In the battle against accidents, injury and loss costs: "It's shaken; not stirred!" To achieve Safety Excellence results...You must: "ROC Your Organization!"

Answers to 'Wiz-dumb' Quiz: NOT! NOT! NOT! NOT! NOT! NOT! NOT! NOT!

ROC YOUR ORGANIZATION!

LARRY L HANSEN

Originally Published as Occupational Health & Safety Online Feature.

***You can instigate Radical Organizational Change
to achieve safety excellence.***

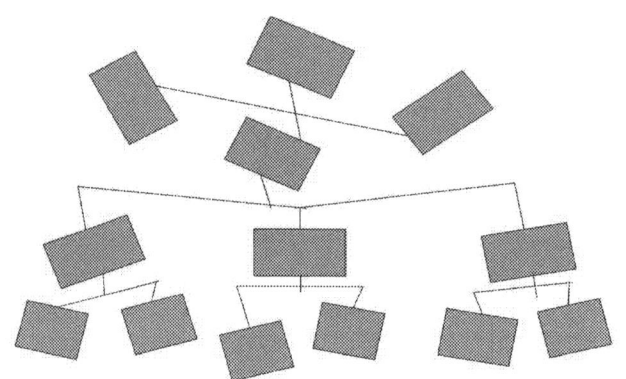

Does this sound familiar? It's a scenario repeated in organizations year after year . . . the dreaded annual planning process! It starts at or about the end of the third quarter--Sept. 30, give or take a week. The CEO returns from the annual "Performance Improvement Strategy Session" with the Board of Directors and calls a Friday morning staff meeting. S/he starts this meeting with one of two very predictable introductions:

"Ladies and gentlemen, the Board is 'very disappointed' with our nine-month results; projections indicate that we won't make our year-end numbers. They've made it perfectly clear that our performance must improve. We've got to ratchet things up through the fourth quarter. Next year they expect a 10 percent improvement across the board with no budget increase over this year's plan! "

Or, perhaps, this:

"Ladies and gentleman, the Board is 'very pleased' with our nine-month results; projections indicate that we'll make our year-end numbers. They applaud all of you for your efforts. However, in view of _____ (you can fill in this blank; the variables are many), they've made it perfectly clear that our results must improve. We've got to ratchet things up through the fourth quarter. Next year, they expect a 10 percent improvement across the board with no budget increase over this year's plan!"

S/he continues: "I'd like all of you to assess your operations, review your variance reports, and bring me a plan that will produce the required improvements. Have your drafts on my desk by Monday morning. Thanks, and I know you're all up to the challenge!"

You return to your cube and ponder: "What's a safety manager to do?" More training, more audits, more meetings, more discipline, more awareness campaigns . . . more of what hasn't worked in the past? And by Monday no less--there goes the weekend.

To make it even worse, deep down you know it's not going to produce the results required; you'll be back in the same situation next year. If only there were an alternative, a way to break free of mediocrity, a way to engage the organization . . . a way to create step-change improvement, a way to attack the real causes of loss in the organization.

A major impediment to achieving excellence is an inability to create sustainable change.

There is. It is a focus on Excellence Strategy. Although there are no quick fixes in business or safety, a focus on "doing right things' can (and does) generate rapid returns.

DEFINING EXCELLENCE

Excellence is not the opposite of mediocrity . . . Excellence is different! Brian Tracy, author of "The Creative Manager," says between you and your ultimate goal lies a rock--a significant impediment that prevents you from achieving your objective. To be successful, you must find a way over, under, around, or through your 'rock.'

In my work with companies striving to become Safety Excellence organizations, I've learned a major impediment to achieving excellence is an inability to create sustainable change. By sustainable, I don't mean the ordinary, run-of-the-mill type of change--but, rather, change of the frame-bending, mind-altering type I called "The ROC"--(Radical Organizational Change) in a Professional Safety article:" *Safety Management: A Call for Revolution."* Communications expert John Drebinger says: "You attain the next level of excellence by changing who you are," and "You change who you are, by changing what you do."

"ROC Your Organization" is all about changing what you do. In book, compact disc, or action deck card form, it provides a collection of non-traditional and somewhat entertaining ways of instigating the kind of change necessary to achieve safety excellence. "ROC Your Organization" is a tool designed to help safety managers and their leaders change 'what and how' safety is done in an organization. The book includes the Safety Excellence Mindset quiz, the World-Class Strategy Model, the Safety Excellence Continuum diagram, and a Safety Excellence Attributes self-assessment. The 52 "Insights on Excellence" (quotes from business leaders and proactive thinkers) and "ROC Initiatives" (organizational change tactics) it contains form a call for action. They encourage practitioners to become change insurgents and take those overt and covert actions requisite to becoming a Safety Excellence organization.

There's a sample of ROCs on the next page:

ROC #15: Go Back to the Future

Go back into the deep, dark corners of the facility and ask six (or more) front-line workers in those areas this question: "Where is the next accident going to happen in this department?" Armed with this information, work with them and the department manager to prevent that future from happening! "Beam me up, Scotty!"

Excellence Requires Proactive Efforts.

ROC #16: Create a Masterpiece

Task five teams to break out and draw detailed pictures of your current safety process (i.e., its design, structure, key characteristics, relationships, major challenges, etc.) After these 'Picassos' have been completed, have each team describe the broad strokes and subtle details captured in their 'works of art.' If a picture is worth a thousand words, odds are these masterpieces will be priceless in their ability to identify problems and target improvement opportunities.

Excellence Requires the Big Picture.

ROC #35: I'll Take a 'Dirty Dozen'

Recruit your cynics, rebels, and radicals; individuals throughout the organization who are commonly considered to be opinionated, sarcastic, and outspoken about the company's safety efforts. People such as these usually have good reasons for bad feelings . . . reasons you need to know about, because more often than not, it is the deeply hidden truth wanting to be heard. Involve these voices in a skunk works initiative designed to produce change and make things better. Challenge them to turn negative thoughts into positive actions. Free their minds, seek their ideas, and involve them in developing positive solutions and leading proactive change. Report on what you learn and accomplish over the next 90 days.

Excellence Requires People Who Care.

SURVIVAL SKILLS FOR THE SAFETY PROFESSIONAL

LARRY L HANSEN

Originally Published May 1997 cover/feature in' Occupational Hazards'

To be successful in the new millennium, says this leading loss prevention expert; safety professionals will need to focus on business value, not compliance fears.

Recently, I was asked to address a Safety and Health Conference on the question "The Safety Profession in the New Millennium . . . Will it Survive or Thrive?" At first I was hesitant to accept that invitation, as I wasn't really sure I had the answer to such a profound question. And secondly, if the answer wasn't a very popular one, I wasn't sure I wanted to be the 'grim reaper' who had to deliver it! However, as I thought further on that subject, it became clear to me that it was in fact a critical question for our profession and its timing was 'exactly right.'. . .I accepted.

My research for that presentation caused me to realize some very important things. I came to recognize that:

1. **The World Has Changed**: The business world is now truly a 'global' community. New technologies, international trade agreements, and the economic unions forged and emerging in the Far East, Europe and Latin America truly pit businesses against global competitors. The truth is, we no longer drive either domestic or foreign cars . . . we drive 'global' cars. We need to overcome the limiting paradox: 'Large world, small minds.'

2. **The Business We Do Has Changed**: In the last 15 years, we have undergone a dramatic transformation of our business base. Some countries are no longer invested in a heavy industry, and a manufacturing economy. Many have transitioned to service industries and a knowledge work economy. Today, many employees work 'in their minds.' Safety in the 'workplace' must be redefined.

3. **The Way We Do Business Has Changed**: Through the leadership of 'B-School' academics, a select group of progressive corporate executives, and enlightened business sages including W. Edwards Deming, and others of the quality movement, business management has dramatically changed its philosophies, principles, and operating practices. We have literally turned our pyramidal organizations 'upside down' . . . yet, in many respects,

The Way We Do 'Safety' Hasn't Changed! Safety in many workplaces continues to employ traditional, 'employee attitude and compliance-based' programs separated from the business process, and, in many cases, pursuing issues directly in conflict with the corporate mission and objectives. This I see as the critical challenge to the survival of the safety profession in the 'now' millennium.

As safety professionals, we are no longer 'salaried' employees. We're 'contingent' employees! Our jobs, our careers, and our futures are contingent upon our organization's ability to achieve operating goals, and produce adequate financial results leading to sustainability. Our employment is only as stable as the success of our corporate endeavors.

With this as a reality, I believe there will be further 'shakeouts' in the safety profession; some will face survival, while others who evolve greater management skills and operational mindsets will achieve . . . thrival! What will drive the 'culling' of the profession, will be the practitioner's ability to respond to one critical question: "What measurable value (real worth) do you contribute to the long-term financial health of your organization?". . . And can you convincingly prove it!

Safety, with its traditional focus, has had difficulty producing satisfactory answers to this question. The future of safety requires new approaches and different thinking. Survival of the profession, I believe, hinges upon its ability to redefine its role, embrace the business process, and become a servant of it, and contributor to it.

Here, then, are seven requisite survival skills for the safety professional in the 'now millennium':

1. **Understand Financial Strategies:** Any function that expects to succeed in the future must be able to demonstrate its financial contribution to the 'bottom line' today. Most safety programs don't attach there very well. The safety professional must develop innovative ways to equate the worth of their effort (accident/loss cost savings) to the financial goals and 'margin measures' of the business. Covey's Law: "If there is no margin, there is no mission."

2. **Support the Business Process**: To 'earn our keep,' the safety function must either contribute to the attainment of greater revenues, or reduce the operating expense of the business greater than their cost. To achieve either, the safety professional must understand the company's strategic plans; market share strategies; competitors; and the tactical plans in place to achieve key objectives. The safety professional must become a supporter of and servant to these objectives, and the business functions tasked to deliver them. The safety professional must become a true partner to all other business center managers in the shared task of turning top dollars (revenue) into bottom dollars . . . margin! We do this by effectively managing the 'middle lines' (loss and expense) . We have great opportunity here, and if we're successful, we all win!

3. **Exhibit Personal Leadership**: The safety professional must be 'technically competent' but more importantly, must function effectively as an organizational 'leader.' The safety professional must have

the ability to impact both people and processes through leadership (power of influence) and persuasion skills. Typically, corporate safety managers are not blessed with large staffs (power of position), hence they must develop the commitment of the people (power of influence) . . . the true leadership quality! The only true measure of a leader is the commitment of his/her followers.

4. **Become a Valued Resource:** If the safety professional's message, programs, and products are not perceived by his or her stakeholders as adding value, they will not be used, and their worth will be diminished and ultimately eliminated. Bottom line -- if we're not 'lined up' with the corporate mission and able to provide support services perceived valuable to it by those calling the plays, we'll be sidelined!

5. **Employ Entrepreneurial Tactics**: What the safety profession currently has to offer, most managers, quite frankly, aren't interested in buying (in some cases, we can't even give it away!) . It's critical that this be changed. A big part of such change requires redesigning our product, re-defining our target customer, providing better packaging, and delivering more effective point-of-sale presentations. The professional must act as an entrepreneur to better define customer wants and deliver innovative solutions to them. The safety professional must focus on:

    •   Marketing - determining customers' 'wants and needs' (a/k/a what they're willing to pay for) ;

    •   R&D - developing products and approaches that continuously redefine and exceed wants!

    •   Sales - presenting the 'product' in enticing ways that respond to needs;

    •   Finance - pricing services to create both real and perceived value; and

    •   Customer Service - delivering all the above when wanted, where wanted, and how wanted to assure customer satisfaction.

6. **Develop Collaborative Skills**: The safety professional needs to build 'strategic alliances' within, and beyond their organization, as true success lies in collaboration and organizational synergy, not control and turfs. The professional must create a 'virtual safety organization' where safety becomes integrated with other key functions, particularly operations, human resources and finance. Safety must value those functions, and it in turn, must value from them. Once common goals, missions, measures, and rewards are created, the synergy needed for ultimate success will become self-generating and self-directing -- Hold on! . . This ride moves fast!

7. **Take the Initiative:** Success in any endeavor is ultimately an outcome of 'personal energy.' The safety professional must expand his or her role beyond its technical limitations. Every position has a title, but a title should never define one's role or confine one's function. Just because we have a job,doesn't mean

we should stop looking for work. To maximize personal success, the safety professional must continually seek ways of growing skills and contributing value to meet the needs of the corporation. That's called initiative, and that's something that is always rewarded in a career! Welcome to Safety in the 'NOW' millennium!

SAFETY IN THE
NEW MILLENNIUM

~

WILL IT SURVIVE . . . OR THRIVE?

Twenty (20) threats (opportunities?) to the Safety profession:

A Focus on Symptoms
You can't manage (change) outcomes...you must deal with causes.

The Era of OSHA - Phase II
Less regulatory clout . . . a kinder, gentler OSHA.

Goaled Mediocrity
A tolerance for' average' doesn't drive good results.

Ignorance -- The 'Segregated Profession'
The safety function is organizationally' exiled.'

Success of the Quality Movement
If' they' can do it . . . why can't we?

Misdirected Energy and Effort
Safety programs are necessary . . . but not sufficient.

Corporate C. O. M. A. (Cost Only Mental Atrophy)
Failure to recognize the difference between' cost and expense.'

Outsourcing
Someone else can always do it easier, faster, cheaper...but generally not better.

Disorganized Labor
Organized labor's ranks and impact continue to shrink.

Compliance Strategy
Setting a' low bar' doesn't lift one to excellence.

Safety As A Political Pawn
Safety as victim of' turf battles' -- HR vs. Finance, etc.

Lack of Leadership
American business is over managed and under led.

Preference for the 'Quick Fix'
Trying to solve a complex problem with the wrong formula.

Reputation---Past, Present…Future?
Past practices have' labeled' the profession.

Better Credentialed Peers
All other things being equal—Credentials Count.

'Bass-ackward' Performance Models
For over fifty years, the obvious answer eludes us.

Career Compression
The safety professional needs to expand their capabilities.

Nontraditional Exposures
Violence, stress, vehicle operations -- new exposures.

The 'Delusion' of Success
Managing incident rates, masks continuance of the problem . . .' COSTS!'

Empowerment
Safety is most transferable to those who would value most -- EMPLOYEES.

"The future belongs to those that create it!"

~ John Graham, President
Graham Communications

A UNIVERSAL MODEL FOR SAFETY X-CELLENCE.

SECTION 2

Barriers to Excellence

*"You can't face new challenges with old methods.
Today, nothing fails like past success."*

~ Stephen Covey

*"The practice of management I've discovered, is filled with myths
that are guaranteed to screw up any organization. The real secret
of an effective leader is to learn how to ignore them."*

~ Jack Stack
The Great Game of Business

TEN COMMON BARRIERS TO SAFETY EXCELLENCE

LARRY L HANSEN

As you strive to improve your safety performance, you most likely will encounter several obstacles common to organizational change. These roadblocks exist in many organizations due to fear, frustration, and doubt that situations can improve or that significant savings can be achieved.

Don't despair! We've seen these before and if some exist in your organization, that's normal! Excellence, however, requires that these obstacles be confronted and overcome. Here are ten common barriers to safety excellence. Think deeply on these and don't let them impede you in your QUEST for Excellence – These are your targets for change!

1. **Lack of executive involvement.** Any initiative that lacks leadership lacks opportunity for success! If senior management isn't involved, it won't happen...do not proceed to Step No. 2!

2. **A pervasive belief that employees are the problem.** Most problems are in the process – management designs and owns the process... be prepared to fix what's really broken.

3. **Lack of cooperation and collaboration.** Teams aren't the answer; TEAMWORK is! Excellence is a cooperative outcome of people focused on common goals, objectives and shared rewards.

4. **Lack of information and enabling technology.** Most organizations fail to solve their safety problems because they can't see them. Truth exists in the data (the reasons beyond the numbers) . Excellence requires that we relentlessly seek to understand both. See Section 14- Active Agenda.

5. **Resistance to modified duty/return to work programs.** Excuses, excuses, excuses! You can't afford not to! Don't allow supervisors to be penny-wise and dollar (often 6-figure) foolish.

6. **A belief that the company is 'a victim' and can't exercise meaningful control.** Another excuse for not wanting to act in the first place...see number 2 above – Stop whining!

7. **A belief that the company is 'different' and that comparable measures just don't apply.** An indicator that the organization is detached from reality – all companies are different, none are invincible!

8. **A belief that the insurance company is taking care of 'the problem.'** The problem has two parts – their job (by contract) is to pay your claims – doing this doesn't solve your problem(s) . Be prepared to deal with the other part – your job, eliminating the sources of loss.

9. **A belief that a safety program will solve the problem.** If you have high workers' compensation costs, you most likely have organizational problems no safety program will fix.

10. **Concern about the amount of management time and effort required to succeed.** This is a valid concern! See #1 above!

"Everything degenerates into work!"

~ Peter Drucker

A

Why Business Under Performs In Safety

Doesn't Know How To Change
- Lacks understanding of strategy; i.e., how to integrate safety.
- Lacks resources and support (internal/external)
- Lacks strategic planning skills

Doesn't Want To Change
- Values
- Leadership
- Complacency
- Easier not to!
- Incentivized elsewhere
- Lacks necessary tools

Doesn't Recognize Alternative Strategies For Change
- Knows only traditional model.
- "Been there -- done that!"
- Stuck in the 60's

Doesn't Perceive Need To Change

Lacks:
- Facts & information
- Vision
- Compelling "Business Case"
- Financial Benefits
- Competitive Profiles

Operations Management

Why Safety Programs Under Perform

Traditional Mindsets & Skillsets
- SOSS
- Turn-off to line managers.
- Emphasis on compliance/programs.
- Reactive efforts - (inspection/Obs.)
- Employee as problem.
- "It's all about me!"

Lacks Business Knowledge
- Lacks understanding and alignment with business vision/mission.
- Lacks understanding of progressive strategies.
- Unable to "integrate" safety into business process.
- Fails to attract management interest.

Ineffectively Positioned
- Not positioned to impact change.
- Reactive "doer" role.
- Staff function placement.
- Internal (political) conflicts (HR Vs. RM Vs. OPNS).
- Detached from financial function. (information)
- Poorly defined responsibility.
- Perceived as the enemy. (PIA)

Safety Function

Why Insurance Safety Services Under Perform

Business Knowledge
- Lack understanding of business process.
- Low capability in advanced strategies
- Little understanding of Management science
- Don't appreciate dynamics of 'CHANGE'
- Focus on activities not business results
- Sub-optimizing measurements

Traditional Strategy
- Emphasis on low level strategies
  - Compliance
  - Programs
  - Technical
- Failure to focus on client needs
- Old answers to new problems
- Ineffective contact points

Insurance Safety Service

Performance Systems
- Driven by activity measures (hrs).
- Consulting = # of Recs./Plans.
- Outside-in solutions (programs)
- 'Call' Emphasis - More is good.
- Internal focus - goals & bonuses
- Service determined by client "size"rather than need/risk.
- FUNDING >- 2% Premium
- Staff not aggressively developed
- The 'Fast Fix'. (programs)
- Disjointed LC & Und efforts
- Under utilizing 'group dynamics"

SECTION 3

Introduction to Safety Excellence

"Excellence is not the opposite of mediocrity...
Excellence is different."

~ Marcus Buckingham
First Break All the Rules

THE SAFETY EXCELLENCE EQUATION

ACCIDENTS, INJURIES, and their financial consequences—called "L. O. S. S." (Lack Of Safety Strategy), all have a common trigger, (employee behavior) and common cause(s) -- (management process and practice). Both of these variables—'person' and 'process'—interact to produce an organization's performance (safe or unsafe), and ultimately, its results --'profitability and sustainability'; hence both must be addressed in a comprehensive loss prevention and management system.

Companies that effectively align roles, design processes, lead people, and forge strong safety cultures that place high value on people, generate superior results, including dramatic reductions in accidents, injuries, and loss costs!

An effective accident prevention and loss management process requires a success formula that focuses on the cultural, organizational, and leadership strategies that impact human performance (what we do), and operational outcomes, (what we get) from an organization. Whether the challenge is to help an organization pull out of a downward performance spiral, or guide it upward toward world-class distinction, the 'Safety Success' equation is the same:

$$\text{SAFETY SUCCESS} = \text{CEO}^{\text{U}}$$

Where:

C =   Culture- Values of the organization.

E =   Elements- The safety process.

O =   Organization – Design and alignment.

u =   'YOU'–the ultimate power of success!

For over 60 years, American business has focused almost exclusively on the 'E' in this equation—the technical Elements of safety, (i.e., Education, Enforcement and Engineering) a/k/a- 'The Safety Program'. For an organization to break through to a higher level of safety performance, it must transition from 'safety as a program (staff administered) to 'safety a process' (line driven and owned) . SAFE must become "How business is done." Success requires a strategic approach which identifies and addresses the impediments that influence people's decisions and actions...the culture (values), management (practices) and organization (design) of the business.

ORGANIZATION

The systems, structures, roles and relationships which align the business to achieve:

PROCESS/PRACTICE

The Human Relations, communication, and consequence delivery systems which shape employee behavior and encourage safe practice and:

CULTURE

The values, vision, mission, and leadership initiatives (decisions, actions, and in-actions) which define: importance of

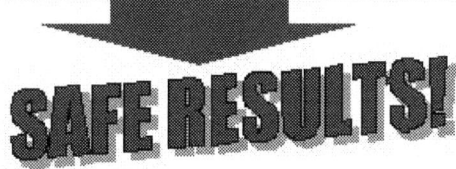

A UNIVERSAL MODEL FOR SAFETY X-CELLENCE.

SECTION 4

Culture

CULTURE - *The unwritten rules that determine what's really important in an organization...What people are willing to 'go to the mat' for!*

"Understand that what we believe precedes policy, procedure, and practice."

~ Max DePree
Past President Steelcase, Inc.

THE 'X'-CELLENCE FACTOR

LARRY L HANSEN

In 1993, as I entered the 25th year of my career and pondered continuing education, two events significantly impacted my professional perspective. The first was a haunting recollection of Dean Wormer's infamous advice to Delta House pledge Flounder in the movie "Animal House:" "Fat, drunk and stupid is no way to go through life son." - it somehow struck me personally. And, the second was a book written by Professor Rich Wokutch, of Virginia Tech. Blacksburg entitled "Worker Protection Japanese Style." The book summarized Professor Wokutch's research, and contained the following trends.

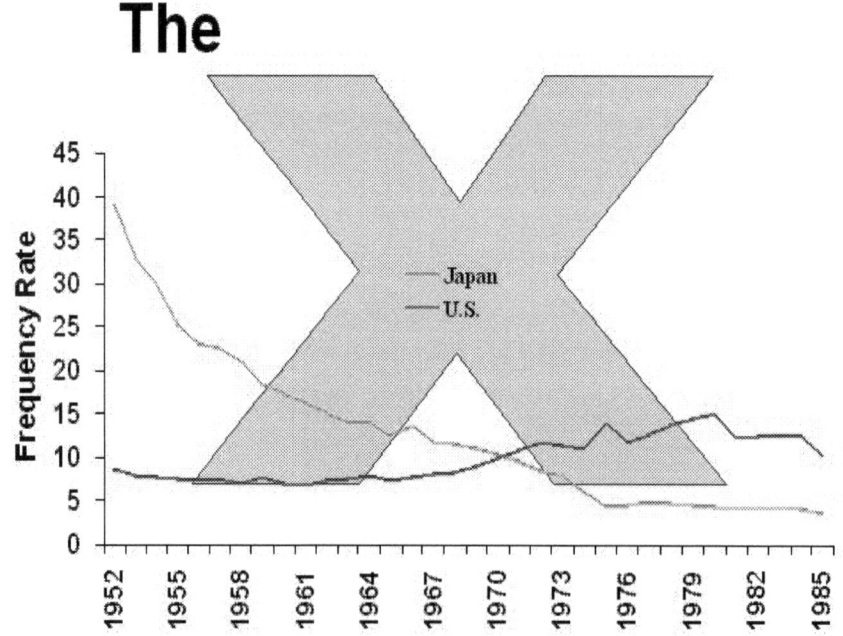

Wokutch "Worker Protection Japanese Style"

~ Rich Wokutch "Worker Protection Japanese Style"

The differences in incident rate trends over time intrigued me, so I read on with interest. Wokutch described his studies of safety programs, practices and outcomes of the Japanese automotive industry, and their comparison to those of the U. S. auto industry. Despite dramatically different (better) results achieved by the Japanese over this 20+ year comparison period, Professor Wokutch went on to emphasize that the Japanese safety activities were basically similar to those commonly practiced by U. S. auto makers. His conclusion from these observations and corresponding data, was that something other than 'safety programs' were the differentiating factor - The 'X' –Factor of safety excellence. Some years later, Don Eckenfelder in his book 'Values-Driven Safety' would reinforce these findings, and make a convincing case for his belief:

"Culture predicts results."

~ Don Eckenfelder

CULTURE IS THE X-FACTOR

~

Culture: *The 'X'-cellence Factor*

Why *do employees violate rules?*

Why *do employees act unsafe?*

Why *do employees take risks and sustain injuries?*

More than 90 percent of all accidents in the workplace are 'triggered' (not caused) by someone doing something they shouldn't have done at a particular time in a particular place. For decades, safety has focused on what people do when accidents occur, i.e., the time, place and risk characteristics of the task, and minimal time on why people do what they do. Yet determining the why (the real causes—what pulls the trigger!) is the most critical part of the accident prevention process, and one which must be thoroughly explored if accidents, injuries and the loss costs they generate are to be successfully minimized.

Employee behaviors stem from their decisions, and their decisions, in turn, are guided by the norms, values, and expectations that shape the culture of their organization. All social groups, employees included, conform to the norms, beliefs and values of their group, and these conforming behaviors frequently have little to do with what's written, ruled, or documented as policy for the organization. In spite of all the controls (safety programs) we implement, we continue to experience a disparity between what we mandate (plan) and what employees ultimately do (behaviors). This, we call "the two safety programs that all companies have"--the one that is 'written and bound' and which exists in procedure manuals, and the one that is 'real' and which exists on the floor in the minds of the workers.

To determine why individuals act in certain ways, we need to examine the real causes of their behavior, the organization's norms, beliefs and values…its culture!

CULTURE shapes **MINDSETS,**
which guide **DECISIONS**
which influence **BEHAVIORS,**
which determine **PERFORMANCE**

which ultimately define results **SAFE** vs. **UNSAFE.**

When viewed separately, each of these factors represents a significant influence on an individual's behavior. When viewed in composite, as dynamic and interdependent success variables, they are powerful shapers of group norms, decisions and performance in an organization; *i.e., why people do what they do.* These 'culture factors' harbor 'the good reasons for poor performance'—the real causes of workplace accidents in an organization!

To succeed, we must manage beyond the paper (programs) and address the culture of the organization; the values, norms and expectations that determine what's really important, and how things gets done. These are factors that forge attitudes, and ultimately shape workers' decisions to operate unguarded equipment (rather than replace the guard), use a ladder that is too short (rather than getting one of proper length), throw 'sand' on an oil spot (rather than repairing the leak), and reach into running machinery to free a jam (rather than shutting down power) . These are the factors that Dr. Deming identified as 'common causes,' —the system -- the cause of 90+ percent of all undesirable process outcomes (including accidents) . These are the factors of performance excellence:

- Safety as a core organizational **Value.**

- A clear safety **Vision and Strategy** as requisite to success.

- The powerful impact of **Leadership Actions** on safe followership.

- **Human Relationships** as fundamental to safe performance.

- A **Communications** architecture to facilitate safe messaging.

- **Integrating** key roles and responsibilities to assure collaboration.

- **Managing Performance** with meaningful **Consequences.**

If poorly managed, the decisions and actions of workers are shaped by the 'culture at hand' and outcomes are unpredictable 'at-risk' and often undesirable. If managed effectively as a system, however, these cultural influences are consistent and actions taken are safe, reliable and have positive impact on the organization's performance and results.

The 'UNIVERSAL MODEL OF SAFETY EXCELLENCE' addresses these seven critical organizational factors and explores those critical inter-relationships, which forge a culture capable of producing: Safety Excellence Results.

THE UNIVERSAL QUESTION:

What Drives Peak Performance?

PEOPLE
Beliefs, Attitudes, and Decisions

BEHAVIORS
(Actions/Inactions)

PERFORMANCE
Human/Organizational Interface

OUTCOMES
Quantity, Quality, Efficiency, SAFETY, Productivity, Delivery, Creativity

RESULTS
Profitability & Sustainability!

Attributes OF SAFETY EXCELLENCE

1. vision/strategy

What managers perceive: An organizational commitment to effectiveness, and injury free operations —doing right things safely the first time.

2. values

What managers believe: An inherent shared intolerance for operational error—making 'safe performance' count!

3. leadership

What managers do: A commitment to action and continuous improvement— making the process ever better and ever safer.

4. collaboration

What managers create: A focus on designing safety into core processes, building collaborative systems, and removing organizational barriers from working relationships.

5. human relations

How managers view and treat people: A strong sense of respect and value for the individual. Building trusting relationships and bonds between employees and the organization.

6. communication

What managers share: A high level of fact giving, information sharing, and encouraged feedback to facilitate safe decision making and enable root cause problem solving.

7. consequence delivery

What managers reward: The building of formal and informal recognition and reinforcement systems and practices that set high value on safe performance.

43

SAFETY MANAGEMENT: A CALL FOR (R)EVOLUTION

LARRY L HANSEN

Originally published as March 1993 cover/feature of' Professional Safety' magazine.

Occupational injury rates are now at a 12-year high and the workers' compensation system, the primary funding source for these losses, is stressed to crisis proportions throughout our nation. 'Traditional' safety programs are not working! Despite these adverse results and contrary to a growing body of data questioning the effectiveness of traditional safety programs, American industry continues to employ these philosophies in the workplace. This article calls for a 'Safety Management (R) evolution' linking safety effectiveness to organizational culture: a process of patterned, predictable change in management values and behaviors. This (R) evolution is identified as a prerequisite to attaining World Class safety performance.

In the aftermath of World War II, W. Edwards Deming challenged American industry to maximize productivity by applying statistical process controls, continuous improvement philosophies and total quality management. However, his challenge went unanswered.

Some 30 years later, Tom Peters presented a new challenge to American management – a challenge for 'organizational excellence' founded in superior customer service and customer perceived satisfaction. Today, a decade after this challenge, Peters reluctantly concludes: "…there are no excellent companies" (Solomon B-8) . He does, however, profess that excellence is now accepted in our management vocabulary and perceives this as part of a new industrial revolution, which will reposition America as a leader in a redefined world marketplace.

American managers have indeed acquired bad habits during the past 40 years. They now, however, recognize that "almost anybody can run a company in boom times because a lot of mistakes are masked by accelerating market conditions. Those same people can sit back and watch their company go down the drain and blame everything but their own lack of action or decision making" (Hoppe) .

These awakenings have created a 'GOOD NEWS-BAD NEWS' scenario:

THE GOOD NEWS IS: An American management revolution is, in fact, underway, with executives across the nation responding to Deming's mandate that "Top management must feel pain and dissatisfaction with past performance and must have the courage to change. They must break out of line, even to the point of exile amongst their peers. There must be a burning desire to transform their style of management" (Deming – Out of Crisis) .

These executives also recognize that such change will take considerable time and energy. Ray Stata of Analog Devices, Inc. concedes, "We are the furthest behind in what I call 'management technology' in my company, in the high tech sector and in all of American industry. Once you awaken to that challenge and decide to do something about it, you may face a 5 to 10 year correction exercise" (Shermerhorn) .

THE BAD NEWS IS: None of this applies to occupational safety!

TRADITIONALISM PERSISTS!

American business continues to embrace traditional safety programs – programs that are isolated from the mainstream of the organization and administered by staff managers who lack the authority and organizational positioning to effect change.

"We've gone from classical management in the 1950's, to human relations in the 1960's, to situational management in the 1970's, to cultural management in the 1980's – all major shifts in philosophy. During these major philosophical management shifts, safety has gone its own way, ignoring reality: most safety programs remain 'classical' in nature; management decides, people follow the rules" (Petersen 47 +) .

Most current programs are based on traditional safety elements, which include: safety directors, safety committees, safety meetings, safety rules, slogans, posters, campaigns and safety incentive programs – all marginally effective. These traditional strategies place safety responsibility with a staff coordinator who is isolated from the line mission and is tasked almost exclusively with inspecting our hazards.

The approach fails to integrate safety into the organization, thereby limiting its ability to identify and resolve management oversights that contribute to accident causation. Activities employed in traditional programs are frequently 'postvention' tracking with 'blood cycles' – being highly visible after severe injuries and, with time, returning to low pre-injury levels. In many organizations, safety is paid for in blood, with little benefit received for the significant amount of safety dollars expended.

"American workers in the private sector suffered 6.8 million occupational injuries and illnesses in 1990. This was an increase of 177, 000 injuries and illnesses over 1989, and results in a 1990 incidence rate of 8.8 per 100 full-time workers, the highest rate since 1979. As in the past, construction and manufacturing, with incident rates of 14.2 and 13.2, respectively, had the highest rates" (1990 Incidence Rate) . This adverse trend is also reflected in injury severity with construction and manufacturing 'lost workday' rates – also at their highest levels, 147.9 and 120.7 days respectively." Loss control specialists cannot confine themselves to safety roles, isolated from others in the organization. Instead, they must work hand in hand with line units, human resource professionals and others to achieve quality and competitiveness goals" (Barge 43) .

TRADITIONAL RESULTS

Research by the National Institute for Occupational Safety and Health (NIOSH) and safety professionals in the private sector continue to document the limited effectiveness of traditional safety approaches. During the 1970s, NIOSH studied management and safety program characteristics in high, low and extremely low incidence rate companies in order to determine which factors most significantly influenced safety results. The findings clearly showed that "commonly prescribed safety practices relating to safety committees, safety rules, accident investigations and reporting, and safety promotions were evident in companies with good safety performance as well as in those with poor safety records. These factors, therefore, are not differentiating ones" ('Safety Program Practices") .

The correlation of safety elements to high and low incidence rate companies indicates that elements typical of traditional programs do not correspond with good safety results. In fact, some elements often perceived as most important, i.e. safety staff and committees, actually have stronger correlations with poor accident results. This suggests that traditional safety organizations (staff-dominated) encourage line managers to abdicate key responsibilities and ignore their critical role in accident prevention. In other words, it's now 'the Safety Director's job! "

CURRENT STRATEGIES

Professor Anthony Veltri, designer of the master's degree program in safety management at Oregon State University, recently identified industry's current focus in safety program structure and strategy. In 1988 and 1989, he surveyed 100 safety professions responsible for their organization's safety programs to determine the predominant strategies being used.

Veltri's surveys examined safety strategies currently employed, needs perceived for the 1990s, and reforms deemed necessary to create added strategic value to the businesses they served. His findings confirmed that traditional safety programs continue to dominate the workplace. Veltri identified three distinct safety strategies, with highly traditional strategies comprising the vast majority:

Reluctant Compliers – 77 percent: This group focuses on regulatory compliance and prefers no substantial safety investment. It pursues traditional inspection activities, with compliance being the key focus of the safety department – not a concern for others. The safety manager's job is to insulate the rest of the organization from compliance problems. This group tends to place blame for compliance demands upon governmental agencies and/or others.

Followers – 16 percent: This group employs creative technical approaches to safety problems and frequently employs modern management and supervisory practices. The tools and programs they utilize have been developed by others, principally leaders in their industry. This group recognizes a need to 'catch up' to the standards set by others, but believes that catching up is simply a matter of doing what others have done before. They fail to recognize that 'catch up' activities cause them to remain outdistanced by the rapid pace set by the leaders.

The Leaders – 7 percent: This group employs truly progressive approaches. They focus on developing distinct functional and organizational capabilities. Their programs add strategic value and build technical capabilities within their organizations. A key characteristic is their commitment to perform significantly better than anyone else in their industry, particularly in areas that senior executives recognize and value highly (Veltri 149 +) .

Both the NIOSH studies and Veltri's surveys confirm that industry's predominant safety strategy remains highly traditional: driven by safety elements, segregated within the organization, and focused on regulatory compliance.

THE IMPACT OF MANAGEMENT

The NIOSH studies also analyzed organization and management factors in the high, low and exceptionally low incidence-rate companies, and found significant correlations between safety results and the core management competencies of an organization. The study concluded: "Management commitment to safety is the major controlling influence in obtaining success, and overall…maximally effective safety programs in industry will depend on those practices that can successfully deal with people variables" ('Safety Program Practices') .

Recent work by Dan Petersen, noted safety management consultant, has also led him to concur with these findings. He concludes, "We believe probably that there's something having to do with the culture and the climate of the organization that makes the whole safety program work. What works in one organization may not work in another" (Sheridan 33 +) .

A recognition that management culture drives safety results is being identified by many industry leaders. John Thirion of Johnson & Johnson acknowledges, "We know that safety is a clear cut barometer of organizational excellence. You cannot have an excellent organization that has a lot of accidents" (Minter 17 +) . Lee Schaller of Dupont Corp. believes, "You have to couple the responsibility for safety with the authority to act. In an industrial enterprise, it is really the line organization that has the authority to act to make the decisions that will effect safety" (Minter 19) . John Maher, Director Of Quality Improvement for Unocal Corp. is convinced that management involvement is critical to safety improvement. Speaking to Unocal's current Safety Improvement Process, he states, "The manuals are not the essence of the process, it is the people, the organization and the attitude. Without that attitude change, that quality commitment to making it happen in the company, a manual is just another book on the shelf" (Minter 48) .

WHAT REALLY CAUSES ACCIDENTS

In addition to these industry leaders, Hank Sarkis, President of the Reliability Group, a management consulting firm in Miami, is also convinced that management culture is the key to safety results. For the past 10 years, Sarkis has collected accident data and organizational information. This data has been statistically analyzed to identify correlations between accident experience and management characteristics. Sarkis now claims his database is credible and is convinced that workplace safety is more a function of organizational issues and management culture than of safety activities and enforcement. In a presentation entitled "What Really Causes Accidents," Sarkis identified the "Top 10 Factors" affecting safety results as:

1. Cheerfulness of Workplace

2. Employee Selection/Placement

3. Procedures – Natural/Awkward

4. Employee Recognition

5. Equipment Safety

6. Pleasant/Stressful Workplace

7. Job Satisfaction

8. Job Challenge

9. Role Clarity

10. Procedures – Safe/Risky

This data again confirms that 'safety elements' have only a limited relationship to good results. Why then do most safety programs continue to be traditional in nature? There are four primary reasons:

1. Lack of Management Preparation. Lestor Therow, Dean of MIT's Sloan School of Management, claims, "To be trained as an American manager is to be trained for a world that is no longer there" (Shermerhorn 516) . His observation is applicable to management skills in general, but is also appropriate to the lack of safety management preparation available in the curriculums of business schools preparing our future industry leaders.

    In his book, Quality, Productivity and Competitive Position, W. Edwards Deming also identifies this weakness in our management development programs, stating, "In fact, anyone could pass with high marks all the regular courses offered in colleges and universities in business statistics and engineering, yet come off with not the faintest idea about how to improve quality, productivity and competitive positioning" (Deming ii) . We can easily add – 'safety' – as well!

2. A 'Microwave Mentality.' To understand accident causation is to recognize that the factors involved in an accident are complex, multi-faceted and difficult to identify. Pressures and priorities facing managers day-to-day, coupled with a lack of technical preparation, creates a propensity for managers to seek 'quick fixes' to safety and health problems. Sarkis recognizes this and has come to believe that "success takes more thinking – failure takes more time" (Sarkis) . Most managers, unfortunately favor the latter.

3. The Lemming Effect. There is a universal belief that the lemming, a small rodent populating the polar regions of the earth, periodically amass in large migrations ultimately leaping to their death in the sea. Even Webster's Ninth New Collegiate Dictionary contains such reference. Researchers, however, confirm no evidence whatsoever concerning such behavior and acclaim it but a myth!

    And so it is that we 'migrate' to traditional safety. Companies faced with an accident problem typically follow what others have done and adopt programs 'off the shelf, ' thus perpetuating traditional safety in the workplace. Bill Cosby once commented, "I don't claim to know the secret of success, but I sure do know the reason for failure – doing what everybody else says to do" (Cosby). And so it is with traditional safety programs.

4. The Fourth Symptom. John Graham, President of Graham Communication, has identified what he calls 'CEO Syndrome' – five symptoms of management ineffectiveness. The 'fourth symptom' is a tendency for CEOs to think in the past. Graham holds that "CEOs perceive their job as getting a company shaped up and running smoothly. Once that is accomplished, they conclude it will continue to run smoothly and that being a CEO is basically a reward for a job well done. They stop the clock at a point in the past when the company was in running order. Problems arise when modern day challenges and management controls created in the past are no longer appropriate" (Graham 6) .

    This is particularly true in safety. Most programs were developed years ago, founded upon principles out of sync with today's challenges and in serious question as to their original creditability.

Unfortunately, these beliefs are deeply rooted in many organizations and heavily impact the safety results they achieve. A company's ability to improve safety performance is directly related to its ability to change its organizational culture – a process both evolutionary and revolutionary, patterned and predictable. This 'Safety Management (R) evolution' is characterized in stages.

STAGE 1 – THE SWAMP
SAFETY WITHOUT ANY MANAGEMENT PROCESS!

These companies are frequently led by the Tyrant-a-Saurus, a management species that has evaded extinction and still survives in some organizations. William Byham, in his book Zapp! – The Lightning of Empowerment identifies these managers as modern-day dragons, who rule with fear and 'Sapp!' employee innovative spirit.

These companies reject responsibility and perceive safety as a burden – a task with no productive value. They accept accidents as a cost of doing business; are autocratic; and have a heavy task focus, with safety frequently compromised to production demands. Their planning process is short-term and reactive; communications are one-way (down) and founded in fear. They employ make-do solutions to equipment and facilities – often unsafe. Minimal employee involvement is allowed in decision making and employee/ management relations are adversarial. It's 'them versus us!'

These companies have high insurance costs driven by both frequency and severity. Their Experience Modification typically exceeds 1.25 (25 percent worse than average). They populate the high risk pools and adversely affect insurance rates for their industry. These companies operate in statutory ignorance, often in violation of recognized safety codes. Employee complaints and whistle blowing occur frequently. These corporations become targets of workplace litigation emanating from major injuries, and frequently make national headlines.

Companies mired in the SWAMP remain there until a Significant Financial Crisis (SFC) occurs – normally, an increase in operational costs or losses so damaging (to profits) that management recognizes a problem and declares, "We need a safety program!" With this, the evolution to Stage II begins.

STAGE II – THE NORM
NATURALLY OCCURRING REACTIVE MANAGEMENT

At this stage, companies implement safety activities without having adequate understanding of their problems or the actions necessary to resolve them. As a consequence, they implement safety programs patterned after 'what others have done,' i.e., assigning a safety director, creating safety committees and challenging them to solve 'the safety problem.' They do not recognize 'the management problem!'

Line managers typically excuse away accidents, i.e., "It was the employee's fault!" They conflict with the safety officer who they perceive to be a nitpicker looking over their shoulders. Line managers do not recognize their key responsibility for safety and normally embrace quick fixes and 'instant-pudding' programs – neither of which work.

Safety programs have high visibility with slogans, campaigns, gimmicks, contests and incentive programs. Managers issue policies, but personally compromise most of them with their behavior,
sending a clear message to employees – 'Read My Lips!' Programs are cyclical. They are implemented, last a short time and then vanish.

Activities focus on inspecting out hazards. This process never identifies problems but rather covers symptoms. Line management never buys into these programs. There is, at best, superficial support. Insurance costs in these organizations are average for their industry. Their Experience Modification is 1.00, plus or minus 25 percent.

Excepting but few – the NORM – is where most companies exist and will remain! For a company to advance to Stage III, it must create radical organizational change (ROC), discarding traditional principles and adopting progressive concepts. These become the record holding companies identified by NIOSH and the leaders identified by Veltri.

STAGE III – WORLD CLASS

In these companies, line management owns and drives safety. Line managers perceive safety as a good business investment with long-term positive returns. Managers believe that accidents are intolerable; they accept no excuses! In these organizations, safety isn't safety; it is organizational effectiveness. The decisions managers make are time-consuming and their planing is long-term, normally three to five years. Responsibilities (both line and staff) are clear. There are shared missions and cooperative efforts.

These companies shun 'just add water' approaches. Their employee relation's policies are humanistic. Employees are empowered and rewarded, often through gain sharing. Communications are open, informal. Feedback is encouraged. Methods to produce safety are built into job descriptions and processes. Results are closely monitored. Causes for variations are identified and rectified.

These organizations are 'quiet!' There are no campaigns, flashing lights, bells or whistles; there are simply results superior to all others. Insurance costs are significantly lower than average, with Experience Modifications less than .75 (at least 25 percent better than average). In these organizations, safety loses identity; there are no 'safety programs.' There are few accidents; there is simply 'excellent management.'

The inability of most companies to attain World-Class performance lies in their inability to create and effectively manage change. They fail to recognize that a business enterprise exists to create change. Peter

Drucker identifies this as one of five survival objectives for an organization: "Every business, to survive, must strive to innovate; that is purposeful, organized action to bring about the new" (Drucker 158). John Graham also identifies an organization's inability to create and manage change as clear indicators of impending business disaster.

Indications of Business Disaster

1. An inability to make anything new or significant happen (create change).

2. The prevalence of a rear view mirror mentality (the past always sets the pace for the future).

3. A run-for-cover approach to crises (a belief that ignoring a problem will cause it to go away).

4. Follow the leader and never take a bold step (belief in maintaining the status quo) (Graham "Five Indicators" 28).

In occupational safety, change of the magnitude needed to produce successful results will only occur when management and the safety professional recognize the need to cast off traditional approaches and force change in corporate culture.

Management must:

- Encourage ideas and reward innovations. Managers must seek out those with the ideas, interests and perseverance to pursue nontraditional paths and turn failures into success, as Art Fry did by transforming a failed adhesive formulation into one of 3M's most successful business products: the Post-It note.

- Demand Superman performance. In his series, "Creative Thinking," Mike Vance directs managers to refocus their values to 'the top line': those central values of fairness, ethics, human relations and organization that determine 'the bottom line' – market share, productivity and profit. A proper focus on the top line will cause bottom lines to fall into place. The message for safety is: What we do – will determine what we achieve. We must do the right things!

- Encourage leadership. An unfortunate reality is: Management does not make a difference in occupational safety and health – or at least has not yet! The American workplace needs leaders with vision, empowerment skills and a willingness to cause change. Max Depree, Chairman of the Board and CEO of Herman Miller, Inc., recognizes that "leaders, in a special way, are liable for what happens in the future rather than what is happening today" (Shermerhorn 205). Industry must identify its 'safety leaders' and encourage them to innovate, experiment and implement those philosophies that will produce positive, lasting results.

- Empower the people. Managers must revisit their working relationships with employees. There is an emerging realization that the next 're-unification' will occur in American management and will link employees with their minds. Empowerment, teamwork and participative management must overcome commonly held beliefs that safety programs fail for one simple reason – employees! This mind set must be replaced with the recognition that employees fail for one simple reason – management! The safety professional, in turn, must recognize and accept an equally challenging role in this revolution. Safety professionals must step up to the challenge that management will place upon them to become, above all, 'managers.'

For too long, safety professionals have hidden behind a Dangerfield Complex. Upon receiving their certification, they wonder when the respect of executive management will follow. (Bielli) It won't! "Participation among top management ranks should not be viewed as a right. It must be earned through responsible performance. When the safety and health professional becomes concerned with promoting the cost-effective use of organizational resources, it will be further empowered through membership among top management ranks" (Burk 46 +).

World-Class effectiveness will only become a reality when all management – executive, operations, line, staff, and supervisory – fully integrate safety responsibility into the organization's mainstream. This will not result from safety programs that are superimposed upon the organization, but only when safety is fully accepted as an integral part of the organization and its mission.

Are we close to achieving this? Unfortunately not! The findings of Veltri, NIOSH and Sarkis confirm a strong attachment to traditional safety. Evaluations by this author further confirm the continuance of this bond.

In September 1991, 24 managers participating in a Nationwide Insurance National Accounts Safety Seminar in Columbus, OH, completed a Safety Management Assessment profiling their organizational 'culture' in safety leadership, vision, integration, communications and human relations. A similar assessment was completed by 118 participants at a March 1993 American Society of Safety Engineers (ASSE) Professional Development Conference in Syracuse, NY.

Both groups determined their predominant culture to be traditional. In the nationwide assessment, 21 percent perceived their safety culture as SWAMP, 71 percent traditional and 8 percent progressive World Class. The ASSE group profiled their predominant cultures as 8 percent SWAMP, 69 percent traditional and 23 percent World Class.

By evidence of these assessments, occupational safety has not yet expressed its readiness to accept those changes necessary to become part of the American management revolution. To be successful, managers and safety professionals must commit to a progressive operational philosophy of "Never leave well enough alone, because the future is too important to let happen by itself" (Primavera Systems). The safety profession needs a (R)evolution!

REFERENCES

Barge, Bruce. "Total Quality: A Risk Management Opportunity," *National Underwriter*, Feb. 17, 1992: 43.

Bielli, Ted Jr. Personal correspondence, Oct. 28, 1991.

Burk, Jerry. "Strategies for the Future of Safety & Health." *Safety & Health Magazine*, Dec. 1991: 46-48.

Cohen, Alexander. "Factors in Successful Occupational Safety Programs." *Journal of Safety Research*. Dec. 1977: 168-178.

Cosby, Bill. *The Bill Cosby Show*.

Deming, W. Edwards. *Out of Crisis*. Cambridge, MA: MIT Center for Advanced Engineering Study, 1982.

Deming, W. Edwards. *Quality, Productivity and Competitive Position*. Cambridge, MA: MIT Center for Advanced Engineering Study, 1982.

Drucker, Peter F. *Technology Management & Society*. New York: Harper & Row Publishers, 1972.

Graham, John. "The Five Indicators of Business Disaster." *National Underwriter*. Sept. 24, 1990: 28.

Graham, John. "Recognizing the Symptoms of "CEO Syndrome". *Central New York Business Jrnl*. Dec. 30, 1991: 6.

Hoppe, Denis W. "CEOs Take Note." *Central New York Business Journal*. Feb. 1991.

Minter, Stephen G. "Creating the Safety Culture." *Occupational Hazards*. Aug.1993: 19-21.

Minter, Stephen G. "Quality & Safety: Unocal's Winning Combination." *Occupational Hazards*. Oct. 1991: 48.

"1990 Incidence Rate Highest Since 1979," *Occupational Hazards*. Jan 1992: 12.

Petersen, Dan. "Safety's Paradigm Shift." *Professional Safety*. Aug. 1991: 47-49.

Primavera Systems, Inc. Advertisement. 1990.

"*Safety Program Practices in Recordholding Plants*," U.S. Dept. of Health, Education and Welfare. Morgantown, WV: Ntl. Institute for Occ. Safety and Health, Div. of Safety Research. March 1979.

Sarkis, Hank. "*What Rally Causes Accidents*." Original research and presentation at Wausau Insurance Safety Excellence Seminar. Canandaigua, NY, June 1990.

Sheridan, Peter J. "The Essential Elements of Safety," *Occupational Hazards*. Feb 19, 1991: 33-36.

Shermerhorn, John R. *Management for Productivity*, 3rd ed. New York John Wiley & Sons.

Solomon, Jolie. "In Search of Results." *Business Today*, March 25, 1991: 3-8.

Veltri, Anthony. "Transforming Safety Strategy & Structure." *Occupational Hazards*. Sept. 1991: 149-152.

DEBUNKING SAFETY MYTHS

LARRY L HANSEN

Originally Published in October 1994 issue of Industrial Safety & Hygiene News
Reprint from Industrial Safety & Hygiene News, October 1994, Vol. 28, No. 10, Chilton Company, Chilton Way,
Radnor, PA 19089.

"True accident causes lie in corporate offices and planning rooms."

It's time to change the lackluster results produced by traditional safety wisdom. To do so, we need to separate myth from reality, and create a new set of core beliefs – a new safety paradigm. Here's what I mean:

Myth: Accidents drive workers' compensation costs!

Reality: Claims drive workers' compensation costs!

People frequently confuse accidents with claims – they're not, in fact, one in the same…one costs a lot of money!

It's necessary for all businesses to have a 'claim deterrent process,' a strategy which goes far beyond accidents or injuries and deals with their ultimate outcome: claims. Claims (the dollar value of accidents) are to a large extent subjective; a matter of employee perception and attitude. Employees' decisions to file claims, lose time, the amount of time lost, their willingness to return and the ultimate degree of residual disability they sustain are all choices employees make based on their perception of the organization and its management values.

Human resource practices offer great opportunities to shape attitudes and slash workers' compensation 'claims.' When we fail to build positive employee relationships, we just fuel the 'claim development process.'

Myth: Traditional safety programs are valid, well founded – they work!

Reality: Traditional safety programs are often based on shallow management support.

Professor Anthony Veltri of Oregon State University conducted a survey to determine the safety strategies most frequently employed in our nation's workplaces. The results clearly identify the predominant strategy (77 percent) to be 'reluctant compliance' - the safety department's job is to shield the line organization from the regulators and to assure statutory compliance.

Alfie Koln, in speaking to industry's focus on quality results via compliance says, "If temporary compliance is the goal of managers, then we just explained the problem with American industry. Temporary is obviously inadequate. As for compliance, quality never comes from mindless obedience." Neither will safety!

Myth: Management 'commitment' is the key to safety success.

Reality: Management 'action' is the sole requisite for safety success.

Talk is cheap; most safety programs are a lot of talk – a/k/a- lip service!

'Commitment' is a passive state of acceptance and can never direct the complex interactions required to improve an organization's safety performance. Only active involvement can overcome the corporate inertia, which impedes an organization from attaining higher levels of safe performance. Commitment without action only produces 'cynicism.' "They watch your feet not your lips," says Tom Peters.

Myth: Poor employee attitudes are the cause of the workers' compensation problem.

Reality: Poor management practices cause all employee attitudes.

All business issues usually get reduced to but two acid tests: 'cost benefit ratios' and 'make or buy decisions.' Without a doubt, employee attitudes are a 'make' decision by managers. Managers don't intentionally hire 'bad attitudes'…they're smarter than that. But, that leaves only one other conclusion: if bad attitudes prevail in the workplace, then they're highly efficient at 'making them.'

Bad attitudes are an issue, but they're not the problem. The problem is their cause, the reasons they exist, and more specifically, the practices, which create them. Hal Rosenbluth, co-author of "The Customer Comes Second," believes that: "Business earns the bad attitudes of its employees." And, J. Michael Crouch, TQM author, says "Employee attitudes are important, but the fact is they are irrelevant until management attitudes are addressed." Michael Shor, President of Health Care First, Inc., is totally correct: "The best loss control program in the world can never make up for lousy employee relations."

Myth: Unsafe employee acts are responsible for 85 percent of all workplace accidents – employees are the problem.

Reality: The process, designed and administered by management, is responsible for 94 percent of all outcomes (including accidents) – management makes most of the mistakes!

Tom Peters speaks of "a blinding flash of the obvious," a phenomena in which obvious facts just don't lead to obvious conclusions. In safety, there's most definitely "blinding flash of the obvious" mentality. When managers are asked "Who's responsible for the production process – planning, organizing, staffing, developing specifications, planning work process, specifying materials, establishing rules, designing layouts, etc. – there's usually no debate that these are management responsibilities. However, when employees are injured from this process, management's typical reaction is: "Careless employees!" Employees sustain injuries…accidents result from the process designed, developed and operated by management.

Myth: Compliance to safety rules assures safe performance – obedience is required.

Reality: Rules can never adequately address the hazard variables inherent in a dynamic organization. 'Thinking' is critical!

Obedience and thinking are at opposite ends of the business spectrum, directly aligned with failure and success. Progressive companies recognize that success is not achieved via rules – employees will follow rules no matter how stupid they are!

In the past, business was run under the premise that managers did the thinking and employees did the 'doing,' no thinking allowed! The new philosophies now call for empowerment, participation and employee involvement. However, this is frequently a ploy. Managers tell employees they want them to participate, give opinions and take part – 'to think' – yet when they do and tell managers what's really wrong (mostly with them and their systems), they are ignored, chastised, labeled as 'not team players'… or down-sized! America's work places do need to be 'reengineered,' but what is really needed is more employee 'head room.'

Myth: Supervisory accident investigations reveal critical facts, which prevent accidents from recurring.

Reality: Supervisory accident investigations rarely identify real accident causes embedded deep in the organization. Recurrence is inevitable.

If we are to believe the findings in accident investigation reports, then the following are the real sources of accidents: 1) Careless employees – 40 percent; 2) Beats me – dunno! (Blank space on report) – 25 percent; and 3) All other causes – 35 percent. Obviously, such conclusions are open to question – but they seldom are! If accident investigations aren't identifying system failures, they're not producing accurate information. Most aren't!

True accident causes rarely lie on the production floor…symptoms do. True accident causes lie in corporate offices and planning rooms, places not generally frequented by safety directors. Managers need to cease reliance on inspecting hazards out of the process and dedicate more effort to designing safety in.

Myth: Safety incentive programs are quick, easy and inexpensive ways to drive safety improvement.

Reality: Safety incentive programs are quick, easy and inexpensive!

Alfie Koln, author of "Rewards as Punishment," identifies key reasons why incentive programs have minimal impact on long-term accident costs:

- First, they're only incentive programs, they don't obligate any change in existing processes or practices;

- Second, incentives ignore reasons, they frequently disguise the real deficiencies and strategic flaws in the organization; and

- Third, they're premised on 'wrong headed' assumptions that accidents are intentional employee acts and that a baseball cap, belt buckle, or $25 savings bond (current value $18.75) will cause them to stop throwing limbs into unguarded machines or falling off of scaffolds.

Myth: To improve safety, an organization must make a significant commitment to employee training.

Reality: To improve safety, an organization must make a significant commitment to fix whatever it is that's really wrong – generally not the employee!

The real sources of accidents are deficient planning, poor organization, unclear goals, lack of vision, vague responsibilities, autocratic direction, lack of vision, vague responsibilities, autocratic direction, lack of employee involvement, conflicting priorities, poor communications and incompetent supervision to name but a few.

When these factors interact and culminate in accidents, management's most frequent response is "we need a training program" – an emphasis on 'people at fault' rather than 'process at fault.' The truth as identified by the late W. Edwards Deming is that management is responsible for most all outcomes of the production system including its volume of human scrap. It's the process that needs fixing 94 percent of the time…not the people.

59

Myth: Safety is an employee benefit issue most effectively handled by committees.

Reality: Safety is a boardroom issue, which can only be positively affected by that group.

Workers' compensation costs have escalated in most industries to be truly a boardroom issue. And yet, how do corporations typically deal with such problems? They create staff/employee committees which lack direction, have limited funding, and which lack the authority to truly impact real organizational causes. The result: meetings every Tuesday of the month whether they're needed or not!

Positive results just can't be produced by 'safety committees,' but they can always be produced by a 'Board of Directors' once they put their minds to it…and they usually only have to meet quarterly!

THOUGHTS ON THE PROFESSION ...AND MAKING A DIFFERENCE

DAN ZAHLIS & LARRY L. HANSEN

Originally Published January 2004 in' Occupational Hazards' magazine

Boy, do we have some good news for you!

A national reader's poll conducted to 'take the pulse' of the profession has just confirmed that job satisfaction among safety practioners has plummeted to an all time low…now hovering in the twenty percentile range.

"Hey, I thought you said you had good news?"

We do! And it's this…If we were Kings and had 'head-lopping' power, we'd fire most of these whiners hunkered down in fear, protecting the status quo, complaining about the very bureaucracies that grant them purpose, and sliding in backwards every week to pick up their paychecks! …And, we just saved a bunch of money on our car insurance!

Almost 80% dissatisfied—that's a big number! What is buried in these statistics, and those that comprise them, (been to a local safety council meeting lately?) is that they complain they aren't supported, aren't listened to, and haven't received their due respect from senior management. Yet, in the next breath, they openly acknowledge that they haven't expanded their knowledge base, explored new strategies, dug-in their heels of conviction, and are fearful of 'pushing back' in their organizations. They simply accept their plight and rationalize their boss's view of them as window dressing, or a necessary expense (for now), and then have the nerve to claim they're under paid!

The problem with most under-performers in any profession, including this one, is the reason for their problem. They're so comfortable making money doing it the way it has always been done that they find it easier to apply old thinking to new problems, and are content in doing more of what hasn't worked before… more of the 'SOSS'- (Same Old Safety Stuff). They would rather continue doing what's 'politically safe,' than stand tall in their convictions, confront their organizations, expose real problems, and advocate (no demand) real change…as that would mean jeopardizing their salaries, and moving out of their personal comfort zones.

Let's rephrase their unhappy and unfortunate 'current reality' this way. Far too many so called practitioners are content (and allowed) to just show up at the office…or plant…or their client's place of business, do meaningless 'stuff', fill out a 'time justification record' to satisfy some corporate quota system, and then leave. However, they'll be back---next day, next week, next month…same time, same place, same 'bat-channel', same problems! They don't actually do much of significance, although they know how to, and could, if it ever were to become necessary. Anyone else on Earth can do their job if they look presentable, keep their mouth shut 95% of the time and can rattle off some antiquated theories, i.e., ('Key Man, 'Iceberg', '3 E's of Safety, Dominos, etc.), and recite verbatim, the titles, sections, paragraphs, and verse of a handful of 'regulatory' minimums. And, they get paid for this. So… "What the heck…only three days till the weekend."

"Try this as an alternative - ROC the boat, take a career risk, dare to make a difference, do the right things, and quit whining…or just quit!"

Now, admittedly, we're not the richest safety professionals on the planet, but we are willing to exchange the financial temptations, creature comforts, and token perks, for freedom of choice, the chance to stand by our convictions, and the opportunity to drive real change in organizational mindsets and processes.

Our recommendations to the '80 percent' identified in this recent poll are:

*   Identify those clients and/or internal customers that see value in your services, and concentrate your time and energy on them. Do so, even if your boss gets pissed…and do so even more if it gets you fired.

*   Make your customers your boss, rather than the Accounting Department, Corporate Rules Trolls, or the corpse that signs your paycheck (it's likely a rubber stamp anyway).

*   Get results or get fired…because it's only results that count, and it's only results that will ultimately free your soul, and enable you to discover the real opportunities to make a difference in this profession, in your life…and in the lives of those entrusted to your care and responsibility!

Continuing to lament: "I really enjoy my work, and I try hard, but 'THEY' won't let me…" keeps you employed, and just might get you promoted, but it will never make a meaningful difference in outcomes, or bring you real happiness. And, if it does, stop whining - you're as much a part of the problem as 'THEY' are - go home and kick the dog!

TEN WAYS TO REALLY MAKE A DIFFERENCE...
OR GET FIRED IN THE PROCESS!

1. **Slash the Employee Training budget!** Stupid people aren't the problem! Double the 'Leadership and Management Development' budgets...enough said! Well, maybe just a word or two more are in order. Training office staff, shipping clerks, and grounds keepers about the hazards of a confined space, fitting a respirator, or how to lock-out process equipment wastes their time, and your money...as do most of the other 11 pre-scheduled compliance topics, which have little bearing on most people's work...yet are obligated to get 'a check' on the corporate audit score.

2. **Discontinue Supervisory Accident Investigations**...they rarely identify the 'root cause' of organizational accidents--unless, of course you're willing to put them behind one-way mirrors, bring in a senior manager line-up, and grant them full immunity. They're supervisors...they're not stupid!

3. **Stick your nose EVERYWHERE it belongs.** Encroach upon the 'turfs' of other functions (sucking out redundancy with a straw), create discomfort with your insurance carrier and brokers (by demanding they do something for those commissions), spend money from one budget account to cover the legitimate needs of another (by fixing problems), and be willing to sacrifice the most sacred cows and long standing bureaucracies of the organization. If your CEO fires you, CONGRATULATIONS! - It worked.

4. **Re-structure your organization.** Require that 'shared ownership' replace 'forced accountability'. Build unified business systems (function to function), and collaborative processes (line and staff) NOT functional departments! We all may not be in the same boat...but we are all in the same ocean. Imagine washing dishes at home. Does anyone have a children's dishwashing department, a husband's dishwashing department and a wife's dishwashing department? ...OR do we just have one process, with one set of tools and equipment to do one task? None of us does anything so complex at the task level that it requires a damned department...or silo...or island...or smoke stack—I think you get the gist.

5. **Eliminate 'Rules Trolls' and the folly they produce.** Rules are made to address 5% of the people (who don't follow them), and they alienate the other 95% (who don't need them). Replace rules with 'values based' process guidelines that delineate systematic methods to be taken by 'PEOPLE' to reduce risk. The phrase "Thou Shalt" shall be reserved for a single purpose – "Thou Shalt refrain from using the phrase Thou Shalt!"

6. **Eliminate, now and forever, the word 'ACCIDENT' from the corporate vocabulary.** The term 'accident' is too commonly perceived (and used) by managers as 'a fortuitous, unintended, unexpected, 'S-happens' event, a/k/a- an excuse. Replace it with the word 'incident or operational error', i.e.; a foreseeable, predictable, and very 'manageable' event manifesting from a series of operational oversights. Now that all the excuses have been eliminated...hold managers accountable for improving their process and minimizing operational error.

63

7. **Grant all employees a 'no approvals required' purchase authority.** Five hundred dollars would be good, a thousand much better! Duct tape and cardboard are nice accoutrements for the shipping dock, but far too much of it is used to 'retro-fix' production processes and improve workstations. If employees can outperform the brightest ergonomists with duct tape and cardboard, imagine what they can do with a 'no red tape' spending authority.

    Note: Allow all employees to trade their authority amongst one another in order to address needs of higher cost, and invite the management team to attend an ongoing lesson in teamwork.

8. **De-lawyer your business.** Run an Accounts Payable printout for all expenses flagged as legal services. If purchasing doesn't categorize expenses this way – force them to start TODAY! Lawyers propagate costs. For every legal dollar spent, there will be more brokered to related service providers, and yet more expended on untracked conflict between the organization and its employees, service providers, vendors and suppliers.

9. **Donate the BINGO game and other counter-productive games of chance to the local retirement home!** Then design meaningful programs that incentivize and reward people based on desired behavior change and achievement of goaled and quantifiable activities, which are designed to produce better results... not the fortuitous chance of someone holding a card with a lucky number ...BINGO!

10. **Goal a 300% increase in your 'Recordable Incident Rate' for the next calendar year.** Don't cause more injuries, demand more reporting! ...and then ignore the OSHA classification and focus on treating the injured person with respect while aggressively trending, analyzing and preventing the root causes of the incidents being reported. In fact, issue an incentive award for every incident reported, irrespective of severity (see item 9 above). And, by the way, **Goal a 30% reduction in workers' compensation costs during the same calendar year...you'll beat it by a mile!**

STEPPING UP TO OPERATIONAL SAFETY EXCELLENCE

LARRY L. HANSEN

Originally Published as May 2005 cover/feature of' Occupational Hazards' magazine

In 1985, I dared to ask one single question that ended my 18-year career as a safety practioner. It also, however, impacted how safety would be managed in companies throughout this country, and marked the beginning of a second more productive career as an Organizational Performance Consultant specializing in 'Preemptive' Risk Management. That question was presented to the profession in March 1993 as Professional Safety's cover story entitled: "Safety Management: A Call for Revolution." Now, some ten years later, it is being asked across five (known) continents impacting the thinking of academics, and practices of many global institutions and organizations. That critical question was, (and remains)..."WHY?"

Inquiring minds want to know:

* **Why**…are 'all industry' LWD Incident Rates only marginally improved, in spite of 30 years of federal regulation and enforcement?

* **Why**… do Workers' Compensation costs continue to escalate in many business segments in spite of these incident rate declines?

* **Why**…do multi-location companies with one centralized safety program have such diverse results across their organizations?

* **Why**… did NIOSH researchers find that companies with better safety efforts had higher accident rates?

* **Why**…did a Department of Energy (DoE) study conclude that sites that invested more (% of budget) in safety incurred higher loss costs?

* **Why,..** in many organizations, is 'safety' managed differently than all other business functions? And most importantly,

- **Why**…did HR executives of the Conference Board cite 'SAFETY' when asked what function could be eliminated due to failure to add value?

These questions frame the bigger question: "If 'safety programs' are a common denominator to organizations that both fail and succeed, what then is the 'X-Factor'…the differentiating variable that separates world-class performance (the best) from the vast majority who struggle to maintain mediocrity (the rest)?

THE 'X'-FACTOR

The answer to these questions in specific, and clarification of the 'X'-Factor (excellence differentiator) in general was provided by Professor Richard Wokutch in 1992. In his book: "Worker Protection, Japanese Style", two important insights emerged. The first was a comparison of United States vs Japan injury frequency trends, which indeed visualized the 'X' - Factor, and the second was his observation, that in spite of the vast difference in results, Japanese safety programs were very much the same as those employed by US firms suggesting that safety programs weren't the differentiating factor—culture was: "Concern for safety and health is integrated into the production system: it supports efforts to promote quality, and productivity. Accidents would severely disrupt production, and therefore must be avoided at all costs. Individual workers and line managers take primary responsibility for ensuring the workplace is safe and healthy. They don't rely as much on safety managers, or governmental regulators as is often the case in the United States."

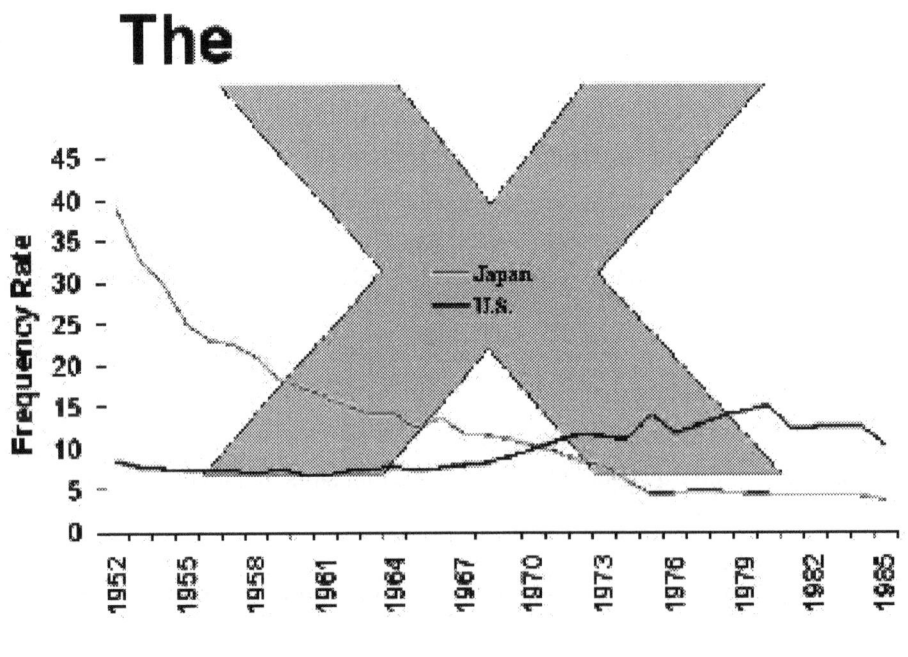

Wokutch "Worker Protection, Japanese Style"

In his October 2003 Occupational Hazards article: "Getting the Culture Right." Don Eckenfelder contends that organizational attitude ultimately determines whether safety initiatives succeed or fail, and proposes three core truths: "1) Culture predicts performance, 2) Culture can be measured, and 3) Nothing is more important than getting the culture right!" (Eckenfelder) The culture of an organization, its basic beliefs and values concerning people, is what drives safety excellence.

Tom Peters and Bob Waterman spent a decade "In Search of Excellence"…attempting to discover what lies at the core of operational excellence. After years of research, they summarized their findings in a simple, yet powerful message to American management: "Figure out your values system!" Values lie at the core of an organization's culture, and are the predictors of, and ultimate determinants of all operational outcomes… safety included.

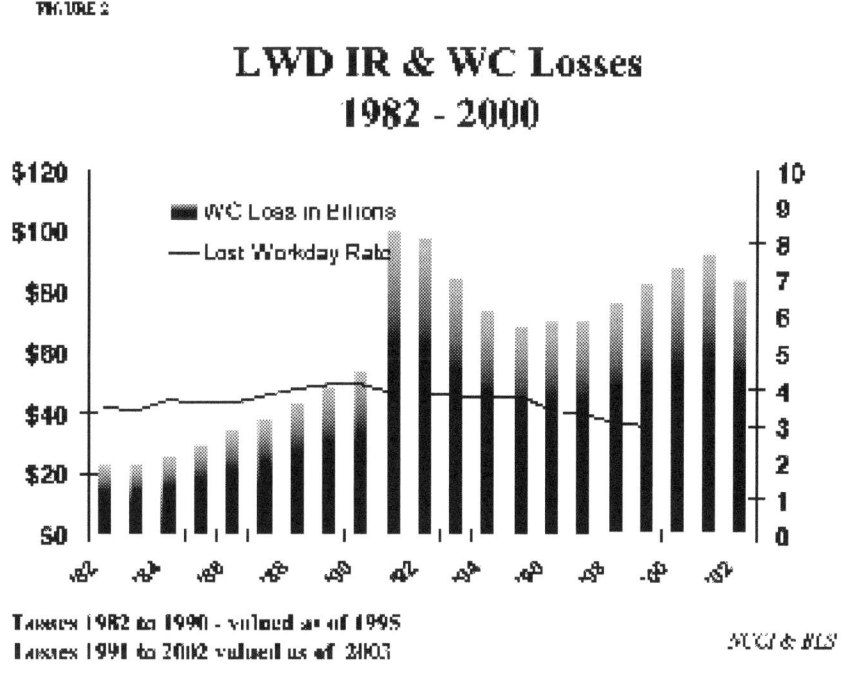

TRADITIONAL STRATEGIES

Stephen Covey warns that: "If we always do what we've always done, we'll always get what we've always got." The obvious question hence is: "What do we do, and what have we got?"

Research by the National Institute for Occupational Safety and Health (NIOSH) initiated in the late 1970's has documented the limited effectiveness of traditional safety approaches in minimizing loss outcomes. When safety effectiveness ratings of a studied population of companies were compared to loss outcomes produced, no significant correlation of 'effort to results' was found. However, when the same population of companies was studied a second time comparing management competencies to loss outcomes, a clear correlation of management effectiveness to low incident rate outcomes was revealed.

A more recent study (safety budget vs loss incurred) at selected sites within the Department of Energy also resulted in unexpected findings…the more significant being (Crites):

1. "Increased investment in a formal safety program did not produce improved safety performance. Distribution of results indicted an inverse relationship, i.e., the greater the safety investment, the higher the level of loss," and

2. Factors having minimal impact were:

    - A shift in safety emphasis

    - Size of the safety budget

    - Degree of hazard

    - Safety rules (quantity or quality), and

    - Safety committees

These, and other similar studies conducted over the past 10 years confirm that management (more than programs) is the major controlling influence in achieving safety excellence, and that overall…maximally effective safety programs in industry will depend on those practices that can successfully deal with people variables. Dan Petersen, concurs with such findings. He has concluded: "We believe probably that there's something having to do with the culture and the climate of the organization that makes the whole safety program work. What works in one organization, may not work in another." (Sheridan). And similarly, D. A. Weaver, a thought leader of the profession 50 years ahead of his time observed: "Excellent organizations frequently achieve exceptional safety results in the absence of any visible safety program, and excellent safety performance cannot be attained in a generally poor organization." His bottom line: "Safety is nothing more than a by-product of doing things right."

THE SAFETY EXCELLENCE EQUATION

More recent Gallup Organization research on high performance companies also identifies values and leadership as key differentiating factors. In "First Break All the Rules," lead researcher and author Marcus Buckingham summarized the key findings of that study: "Excellence is not the opposite of mediocrity… Excellence is Different." Excellence is not generated by more of the same, only faster, quicker, harder; but rather by re-focusing on the drivers of high performance, culture (values) and leadership (practices). Relative to 'safety', this would suggest that traditional safety elements (programs), although valid and necessary, are alone, not sufficient to achieve safety excellence in an organization. These elements need be empowered by the culture of the organization. There is a 'Safety Excellence' equation that applies to all organizations, and it is:

$$\text{SAFETY EXCELLENCE} = CEOu$$

Where:

C = the **CULTURE** of the organization (Values)

E = the **ELEMENTS** of the safety (Program), and

O = the **ORGANIZATION'S** safety performance (Systems).

u = the **LEADERSHIP** (Actions of Executives and Champions)

The Safety profession has spent over 50 years perfecting the 'E' - Elements of safety; it's now time to move forward, and focus on the enablers of excellence—the 'C'- Culture and 'O' - Organization of the business.

When conceived ten years ago, the safety excellence continuum model proposed to define and diagnose culture improvement opportunities, consisted of only three levels, and two requisite 'step-changes'. It is now evident that a fourth 'high-end' performance level, and a third mindset change is requisite to attaining true World-Class distinction. The revised excellence model now consists of four 'step-change' performance levels: SWAMP – NORM – EXCELLENCE and WORLD-CLASS. For those familiar with the original work, this revision will reinforce original concepts. For those more recently entering the profession, this model serves as a strategy beacon to guide efforts toward world-class results. In "Good to Great," Jim Collins and his research team concluded that: "The first step of Leadership is not visioning, but rather confronting the brutal facts." (Collins) Following are the patterned management practices and predominant cultural beliefs that define the current reality of safety, and which must be confronted at each level of the 'step-change' journey to World-Class safety in an organization.

STAGE I – THE S.W.A.M.P.

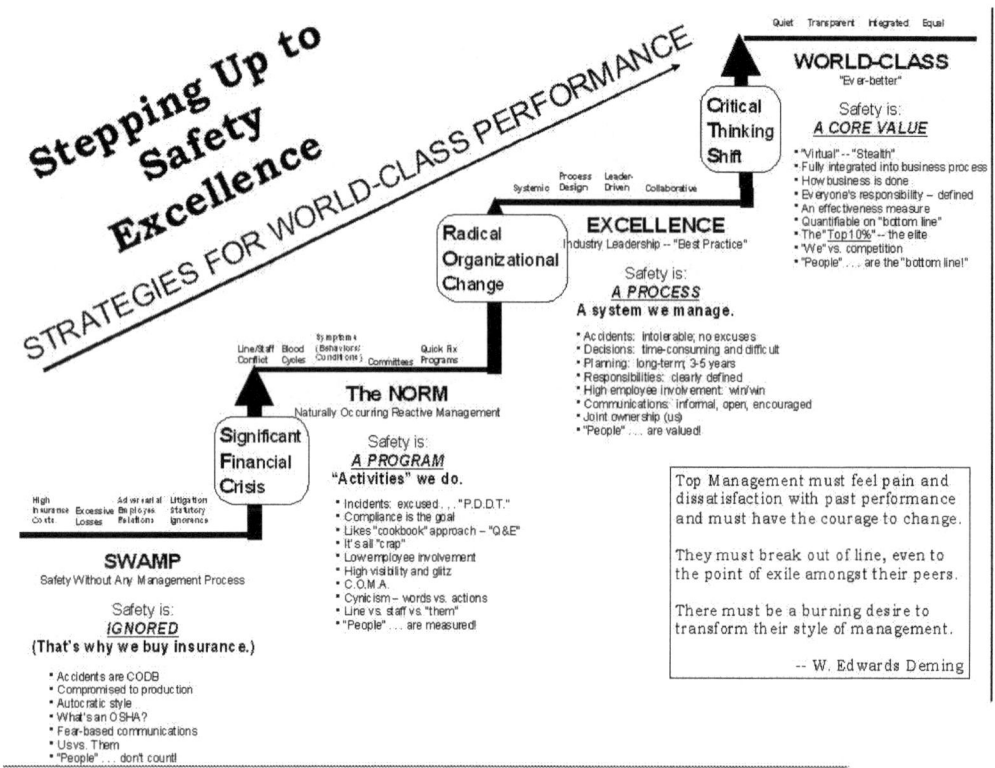

(SAFETY WITHOUT ANY MANAGEMENT PROCESS)
Safety is' Unmanaged' - Safety is ignored!

COSTS ARE THE PROBLEM!

Organizations mired in the SWAMP are frequently managed by the 'Tyrant-a-Saurus Wrecks' a management species that has evaded extinction in many organizations. These companies reject responsibility and perceive safety as a burden placed upon them by regulators, the insurance industry or the labor organization, a task with no productive value. They accept accidents as an unavoidable cost of doing business, are autocratic, and have a heavy 'production focus,' with safety frequently compromised to quota and/or delivery schedules. People are viewed as expendable resources. Their planning is short-term and reactive; communications are one way (down), and founded in mandates of fear. They employ 'make-do' solutions to equipment and facilities problems, often leaving them unsafe. Minimal employee involvement is allowed in the process and labor/management relations are often at odds concerning safety and adversarial on most everything else. It's always a case of 'them vs us!'

These companies have high insurance costs driven by both frequency and severity. Their Experience Modification typically exceeds 1.25 (25 % debit surcharge). They populate the high-risk pools, and adversely affect the insurance rates for their industry classification. These companies operate in statutory ignorance, often in violation of recognized codes and regulations. Employee complaints and whistle blowing occur frequently. They are targets of labor lawsuits and workplace litigation emanating from injuries, which frequently make national headlines. Companies mired in the SWAMP remain there until a Significant Financial Crisis (SFC) occurs, which can be either a single catastrophic event, or a cumulative increase in loss costs so significant as to impact profits, and threaten the CFO's or CEO's position, hence forcing senior manglement (not a typo) to acknowledge a problem and declare: "We need a safety program!" It is with this impetus that evolution to Stage II the 'N.O.R.M.' begins.

STAGE II – THE N.O.R.M.

(NATURALLY OCCURRING REACTIVE MANAGEMENT)
Safety is' Mis-managed' – Safety is a Program!

PEOPLE ARE THE PROBLEM!

Because the decision to act was driven by cost and ignorance rather than an understanding of real causes, the NORM is typically christened with the 'kiss of death'—the hiring of a Safety Director! This is a typical move as management believes people are the problem, hence the natural answer is to hire someone to 'fix them'… not us!

At this stage, companies implement safety programs without having adequate understanding of the problems or the actions necessary to resolve them. They implement programs patterned after what others have done, i.e., create committees, establish rules, implement training, and enforce progressive disciplinary policies. None of these proves effective, as they are answers that do not address the problem...the management problem.

Line managers typically excuse away accidents as employee carelessness. They are in conflict with the safety officer who they perceive to be a nitpicker impeding their real job…to get product out the door! Line supervisors do not accept responsibility for the safety and health of the people assigned to their units, and embrace 'quick fix' programs which have minimal impact, as employees see through the ploys and 'blow them off'.'

Safety campaigns have high visibility, with slogans, contests, gimmicks, and incentive programs. Managers issue rules and more rules, but frequently compromise them in their own day-to-day behavior, sending a clear message to employees: "Read my lips…" Efforts are cyclical as they follow blood cycles--injuries occur, pressure applied, injuries reduce; pressure removed.

Activities focus on inspecting out hazards and observing and disciplining out unsafe work practices. This process fails to identify core problems, and only addresses surface symptoms. Line managers 'do' safety but

71

don't 'buy into' safety. Insurance costs in these organizations show some improvement but plateau at or about industry norms. Experience Modifications hover around 1.00 plus or minus 25 percent…this varies year to year.

The NORM is where many companies exist, and where most will remain. For an organization to advance onward to Stage III…EXCELLENCE, they must undergo a 'Radical Organizational Change' (ROC), discarding traditional beliefs and approaches, and adopting a more progressive mindset on systemic cause and correction. These become the Excellence companies.

STAGE III – EXCELLENCE

(SAFETY EXCELS TO THE TOP QUARTILE)
Safety is' Managed' - Safety is Integrated.

PROCESS IS THE OPPORTUNITY

In excellence companies, safety is less scheduled and more systemic. Efforts are dedicated to building collaborative systems and cooperative partnerships that integrate safety into core business processes. There are few, if any, safety rules, safety meetings, safety audits, safety training, safety metrics, and least of all safety committees. The objectives of such activities are integrated into operational procedures. In place of separate 'safety' activities, there are normally held operations meetings (that include, and often start with safety), there are standard operating procedures and training (that include safety), there are problem seeing and solving sessions (that address safety), and there are manager meetings to address on-going performance improvement opportunities (that include safety).

These organizations are well schooled in TQM concepts, progressive management, principles, and modern leadership practices. Accidents are rare events, and when they occur they are addressed quickly and effectively…at their root cause level. Labor relations are healthy with many of these companies listed on recognized business lists, i.e., "Best 100 Companies to Work For" and/or publicized in business trades, B-school case studies, and management journals. Accident costs are low, and Experience Modifications evidence a downward (credit) trend to .75 or better (at least 25% better than industry average). For these companies, safety pays dividends and adds to the bottom line. Many in this group have transformed their safety function from a cost center, to a profit center in recognition of its ability to make margin contribution and create shareholder value. Excellence companies face one additional 'mind shift' on the journey to becoming a true World-Class safety organization. This final 'step-change' involves a Critical Thinking Shift (CTS) wherein safety is no longer perceived as a technical and/or managerial issue, but as a core value critical to business success. Safety in World-Class organizations is cultural, an issue of leadership values-- 'Safe is how business is done'.

STAGE IV – WORLD-CLASS

(SAFETY AT THE TOP)
Safety is' Non-managed'...Safety is led!

CULTURE IS THE SOLUTION!

In a short, but powerful statement, Peter Drucker summed up the cumulative insight of his five century career as this planet's most influential thinker on management practice: " With 50 years of hard evidence at hand, it's awful hard to 'slough off' the truth...It's all about PEOPLE!" (Drucker) One of the most distinguishing features of World-Class safety organizations is that 'shared ownership' by all replaces 'forced accountability' by few. Line managers accept primary responsibility for leading safe operations, and employees actively contribute to and cooperate with the process because it is founded upon shared values not imposed rules. Senior managers place a high value on the health and well being of people, believe that accidents are unnecessary, intolerable, and preventable...and let those thoughts be known throughout the organization. In World-Class organizations, safety is a measure of operational effectiveness...a key metric of strategic business success or failure. The decisions managers make are time-consuming and planning is formal and strategic. Responsibilities for both line and staff managers are clearly defined and aligned with collaborative partnerships replacing inter-function conflict.

These companies shun quick fixes, knowing quite well they won't work. Their employee relations policies and manager practices are employee centered and humanistic. Employees are empowered and rewarded, often through gain-sharing arrangements. Communications are open and informal. Feedback is encouraged, and the grapevine is more constructive than destructive. Methods to produce safety are built into job briefs and standard operating procedures. Results are closely measured and monitored. Causes for variations are identified and rectified, and there is a predominance of reinforcement over discipline in this process.

There are no flag-waving campaigns, stump speeches, or bells and whistles; there are simply 'good business' practices that produce superior results. Insurance costs and retained losses are low relative to the size and scope of operations. Experience Modifications are among the best in class, and hover historically below .50 (50% credit) or better than the industry. In these organizations, safety loses its identity; there are no 'safety programs'. There are few, if any, accidents. There is simply 'excellent leadership'. As Peter Senge asserts: "Mastery is invisible."

World-Class performance will only become a reality in an organization when all managers; executive, operations, line and staff fully integrate safety into the organization's mainstream value system, policies, and practices. This will not result from safety programs superimposed upon the organization, but only when safety is fully accepted as integral to the organization's mission, and as a strategy critical to the success of business objectives.

73

For those organizations willing to commit proactive leadership, and willing to refocus efforts (it doesn't cost any more money), World-Class safety is attainable, NOW. Peter Drucker's observation, however, on the requisites of business success applies in all organizations: "All theory degenerates into work." Given an opportunity to pursue transformational change (in mindset and strategy), an unfortunate reality is, most defer to Coderre's (Paul) Law of Least Resistance: "Given the opportunity to do nothing, most will." They prefer to employ L.A.M.E. (Lazy – Antiquated – Mediocre – Externally focused) excuses for substandard performance. To these organizations, the soft warm ooze of the SWAMP is too comfortable (numbing), and the status quo of the NORM too familiar (easy). They opt for the more common alternatives (and costs) of mediocrity:

•   Increase the annual Workers' Compensation premium budget.

•   Add legal and claim administration staff.

•   Blame the government, the Union, El Nino, and their useless brother-in-law.

•   Set higher production quotes to offset loss costs, and

•   Lower the bar on margin projections!

World-Class safety is a journey available to all, taken by some, and completed by an elite few.

Watch that first step…
it's a big one…and potentially a very profitable one!

REFERENCES

Arthur D. Little Survey, "Green Wall Between Environmental and Business Staffs Blocks Successful Environmental Management". *Professional Safety*, August 1996

Cohen, Alexander, "Factors in Successful Occupational Safety Programs", *Journal of Safety Research*, December 1977.

Cohen, Harvey H. and Cleveland Robert J., "Safety Program Practices in Record-Holding Plants", *Professional Safety,* March 1983.

Collins, Jim, "Good to Great", *Harper Business Books*, NY, NY 2001

Crites, Thomas, R. "Reconsidering the Costs and Benefits of a Formal Safety Program", *Professional Safety*, December 1995.

Buckingham, Marcus and Coffman, Curt, "First Break All the Rules", *Simon and Schuster*, NY, NY 1999.

Eckenfelder, Donald J. "VALUES-Driven Safety", Government Institutes, Inc., Rockville, MD, 1996

Eckenfelder, Donald J., "Getting the Culture Right", *Occupational Hazards*, October 2003.

Hansen, Larry L., "Safety Management: A Call for Revolution", *Professional Safety*, March 1993.

Minter, Stephen G. "Creating the Safety Culture", *Occupational Hazards*, August 1991.

NOVA Documentary "Peter Drucker – An Intellectual Journey", WGBH TV, Boston MA June 2004,

Peters, Tom and Waterman, Robert, "In Search of Excellence",

Planek, Thomas W. and Fearn, Kevin T., "Reevaluating Occupational Safety Priorities: 1966 to 1992", *Professional Safety*, October 1993.

Sarkis, Hank. "What Really Causes Accidents", Presentation at Wausau Insurance Safety Management Seminar, Canandaigua, New York, June 1990.

Sheridan, Peter J. "The Essential Elements of Safety", *Occupational Hazards*, February, 1991

Veltri, Anthony, "Transforming Safety Strategy and Structure", *Occupational Hazards*, Sept, 1991

Wokutch, Richard E., Worker Protection, Japanese Style; *Occupational Safety and Health in the Auto Industry*, ILR Press, Cornell University Ithaca, New York, 1992.

A UNIVERSAL MODEL FOR SAFETY X-CELLENCE.

Wokutch, Richard E., "Myths of the Japanese Factory", *Journal of Commerce*, August 26, 1992

SECTION 5

Strategy – Vision/Mission

"The future ought to be planned and managed, because it's going to happen anyway."

~ Phillip Crosby

"The key point is, whatever generates a particular result, good or bad, occurs far in advance of the result itself."

~ William Lareau

The Universal Model of Safety Excellence

Safety Leadership

Strategy

A QUESTION OF EXCELLENCE:

STRATEGY

Have we made our expectations for safe operations perfectly clear to all in the organization, and have we effectively planned how we will achieve our vision?

THE PLAN

In the beginning there was the plan.
And from it, came forth the assumptions.
And the assumptions were without form.
And the plan was without substance,
And darkness fell upon the face of the workers.
And they spake unto their Supervisors, saying:
"It is a pot of crap, and it stinketh."
And the Supervisors went forth to the Managers and sayeth unto them:
"It is a pile of dung, such that none may abide by its odor."
And the managers went unto the Superintendent
And sayeth unto him:
"It contains excrement, and it is very strong
such that none may abide by it."
And the Superintendent went unto the Operations Director
And sayeth unto him:
"It is a vessel of fertilizer, and none may abide by its strength."
And the Director went unto the Vice President and sayeth:
"It contains that which aids plant growth, and is very strong."
And the Vice President went unto the senior Vice President
And sayeth to her:
"It reeks of promise to promote growth, and it is powerful."
And the Senior Vice President went unto the president and
Sayeth unto him,
"This powerful new plan will actively generate growth
and efficiency of the company and of the business in general."
And the President looked upon the plan and saw that it was good...
AND THE PLAN BECAME POLICY!

-Unknown (but very insightful) Author

TOP TEN REASONS FOR STRATEGIC PLANNING

1. Because management expects us to achieve greater results which can only be achieved through change--a strategic process.

2. Because strategic plans are a continuous improvement process, and their success increases exponentially with each subsequent year following implementation.

3. Because it defeats the two most prominent frustrations of safety initiatives: the accident (blood) cycle, and the performance plateau (flat lining).

4. Because it's consistent with modern management philosophy. We can't afford to be the last dinosaur in a world that's increasingly embracing change. To do so would discredit, devalue, and isolate the safety function.

5. Because it raises employee commitment by increasing their sense of ownership, understanding of mission, and participation in the company's vision. This improves performance in numerous other ways.

6. Because it builds cooperation, teaming, and collaborative skills among employees, through functions, and across hierarchical boundaries.

7. Because it can generate the momentum to propel an organization into other related successes. (Nothing succeeds like success!)

8. Because it presents an opportunity for safety to assume a leadership role in driving organizational performance and cultural growth.

9. Because it improves the perception and credibility of safety, both upstairs and downstairs.

10. Because it works. Visionary companies significantly out-perform their competitors.

THE ARCHITECTURE OF SAFETY EXCELLENCE

LARRY L. HANSEN

Originally Published as May 2000 cover/feature of' Professional Safety' magazine

There's a growing current of dissension within our ranks that, if unchecked, will impede rather than advance the true objectives of our profession. One need only read the "Letters to the Editor" or "Readers' Pulse" columns of leading professional journals to fully appreciate the extent of this polarization. Faction leaders are digging in and fortifying their positions in defense of their 'one best way' of attaining safety excellence. It's interesting to watch the innuendoes and, at times, direct shots being exchanged between the various camps; (i.e., compliance vs. programs, education vs. engineering, technical vs. management, and behavioral versus cultural). It has become clear, there is little consensus on what need be done to achieve safety excellence. The result of these clashes is a growing state of confusion and frustration within the profession. The demands for greater results placed upon practitioners have increased while the pathways to success have become less clear. To succeed, we must broaden our understanding and soften our resistance to change.

In The 59 Second Employee: How to Keep One Second Ahead of Your One Minute Manager, Rae Andre and Peter Ward of Northeastern University, tell the anecdotal tale of an isolated civilization who toiled for decades to discover the ideal form of social government. After much debate and many years of failed experimentation, the wise elders ceased their efforts and, adopted "The Law of the TANOBWAY" — recognition that There Ain't No One Best WAY! (Andre)

It's critical that safety practitioners recognize that success isn't an 'or' issue (one strategy or another), but rather an 'and' issue (one strategy and another ... and another, and another!) Safety excellence isn't the result of a singular strategy. There are no universal answers.

Peak safety performance results from multiple strategies designed and applied across a broad spectrum of issues and risk factors within an organization. Safety excellence is the outcome of a strategy continuum; a strategic architecture, which addresses the regulatory, technical, engineering, organizational, behavioral, managerial and cultural aspects of an organization.

To put safety excellence into perspective, this paper will:

1.   Construct an Architecture of Safety Excellence (show what excellence looks like);

2.   Identify the components of safety excellence (identify the excellence strategies); and

3.   Define the process of attaining safety excellence (outline the steps, sequence and key linkages).

The pursuit of safety excellence requires that we first address the most critical question of our profession: "Why do accidents occur in the workplace?"

And the answer has become abundantly clear: 'At-risk behaviors' — what people do! Behavior isn't the next level of safety strategy; it's the ultimate level. Behavior is the critical element, which must be addressed to achieve safety excellence! Seventy years of research and observation ranging from Heinrich's early hypothesis in the 1930s to DuPont's time tested successes of today confirm that unsafe behaviors are involved in most all accident occurrences — involved in ... not the causes! The core question remains: Why do employees do what they do, act unsafely, and have accidents?…what 'causes' at-risk behavior?

The vast majority of managers fail to seek the true answers to this critical question. Instead, they rely comfortably on the all too common excuses of: employee carelessness, inattentiveness, disregard for procedure, and laziness (i.e., employee as the problem). This thinking (or lack of) presents the greatest obstacle to safety success. "An organization will never improve its process, if it believes its people are the problem!" (Manule)

The harsh reality concerning safety is that poor performance has good reasons, most of which are inherent in the planning, design, implementation, maintenance, administration and modification of the process — not individuals! Only through the elimination of these process causes (the good reasons for poor performance) can an organization attain safety excellence.

Getting to these good reasons requires a comprehensive change strategy; a strategy which addresses both process and people. Rad Smith, co-author of the QS 9000 quality standards, identifies three levels of change that can be pursued in an organization; each having a progressively greater impact on operational outcomes and results. These are:

Level 1: Corrective Change - fix what's broken — the most common type of change;

Level 2: Continuous Change - improve what is — the most accepted type of change; and

Level 3: Creative (Innovative) Change - doing something totally different — the most profitable type of change! (Paton)

As we build the following seven strategies into an architecture of excellence, keep these change levels in mind. Try to assess your organization's current position, and define the change level and target strategies needed for greater success.

SAFETY PROGRAM STRATEGY (Attitudes, Awareness & Training): 'hink safe1'

The first foundational strategy is that of the safety program. This strategy is based on the premise that safety results will improve by changing the attitudes of employees. This strategy attempts to improve employee safety awareness through policies, procedures, meetings, training, and disciplinary policies. Tactics most frequently employed under this strategy include development of manuals, procedures, rules, committees, policy statements, orientation, training, retraining, remedial training, and ultimately progressive disciplinary programs. Research on training effectiveness, however, has confirmed limited impact on accident rates and costs. A comprehensive study by the U. S. Dept. of Energy on selected sites actually confirmed an inverse relationship (Crites). And, most recently, the National Institute of Occupational Safety and Health (NIOSH) concluded from it's Summit on Safety Training that there is no clear research confirming the effectiveness of training on safe performance.

COMPLIANCE STRATEGY: "You will be safe … or else!"

The second (and legally necessary) strategy is that of regulatory compliance. This strategy is based on the premise that safety results will improve by changing the level of statutory compliance in an organization. This strategy focuses on changes in conditions, facilities, equipment and the work environment in accordance with minimum regulatory requirements. Tactics most frequently pursued under this enforcement strategy include facility inspections, compliance audits, walk-throughs, and programs addressing minimum requirements and action levels subject to citations, fines and penalties.

TECHNICAL STRATEGY: "It's cheaper to bend steel than backs."

The third safety strategy is that of 'technofix' or engineering strategy. This strategy is based on the premise that safety results will improve by changing the level of safety engineering and physical safeguarding in the work place. This strategy emphasizes automation, ergonomics, work methods, workflows, worker/machine interfaces, mechanical advantage, safeguarding, and process design. Tactics frequently pursued under this engineering strategy include ergonomic task assessments, workstation redesign, workflow analysis, ergonomic devices, tool design, and engineering safety into new processes and/or retrofit safeguarding on the shop floor.

> *"If you always do what you've always done, you'll always get, what you've always got."*
>
> -Stephen Covey

These three strategies, in composite, form what is commonly referred to as 'traditional safety' (Three 'E's' of Safety: Education, Enforcement, and Engineering). Dave Johnson in his Industrial Safety & Hygiene News' 1998 White Paper confirmed the continued dominance of these traditional strategies in today's industry practice: (Johnson)

Strategy	Reported Usage
• Program	- 81 percent (Education)
• Compliance	- 74 percent (Enforcement)
• Technical	- 75 percent (Engineering)

Unfortunately, this continued emphasis on traditional strategy hasn't had significant impact on national incident rates or workers' compensation costs over time. Plotting Bureau of Labor Statistics (BLS) national all industry incident rates and National Council of Workers' Compensation Insurance (NCCI) annual workers' compensation costs shows that maintaining a flat (or slightly declining) incident rate progressively drives costs upward!

Safety programs (awareness/training) educate workers, but only have minimal impact on safe work behaviors. Compliance strategies keep an organization legal but don't necessarily lower loss costs. And, technical strategies, although based on sound engineering principles, are often limited in their ability to retro fix obstacles and impediments that can't be engineered out of the process, or existing facilities and equipment.

These three strategies represent the predominant and current state of the art (and science) of safety in most organizations, i.e., much emphasis placed on training, enforcement and engineering, resulting in flat incident rates and escalating workers' compensation costs. Bottom line, these are not wrong things; they are just not sufficient for achieving better results and lower loss costs. As the earlier referenced ISHN Readers Poll confirmed, we are highly efficient (doing things right), but minimally effective (not doing the right things).

Safety Excellence organizations have recognized the need to pursue Level 3 change in safety strategy — doing totally different things. They have shifted from staff administered, antecedent-driven programs comprised of rules, policy, SOPs, and regulations to line owned, consequence-driven management practices and processes. They have embraced the truth concerning safety excellence, offered (yet still not uniformly accepted) by D. A. Weaver in the 1960s - this truth:

"Excellent organizations frequently achieve exceptional safety results in the absence of any visible safety program, while...excellent safety performance cannot be attained in a generally poorly managed organization." (Weaver)

Weaver's premise, I believe, has finally come of age; "Safety really is nothing more than a by-product of doing right things right." (Weaver) Safety is embedded in the business process.

Guided by this truth, world-class organizations are bridging the safety performance gap by setting in place a second set of organizational strategies founded upon values, which forge a safety culture. Building off this second foundation, progressive organizations are constructing the organizational strategies critical to success: Culture (values), Organization (structure), Performance Leadership (consequence delivery), and Organizational Behavior strategies, which link with, and enable the traditional strategies to work successfully.

SAFETY CULTURE: "You can't simply manage your way out of the way you behave." (Covey)

The fourth and foundational strategy of excellence is that of cultural safety. This strategy is based on the premise that safety results will improve if an organization identifies, assesses, and strengthens its values and leadership of safety.

> *"As I grow older, I pay less attention to what men say. I just watch what they do."*
>
> — Andrew Carnegie (Quotes)

Safety culture deals with the 'unwritten rules' (clarified by action) that determine if safety really is important in an organization. Safety culture is forged by what executives do (their decisions and actions) more than by what executives say (their policies and proclamations). Tactics most commonly pursued to strengthen safety culture in organizations are: visioning sessions, mission and purpose definition, and values clarification ... and, above all, commitment to high visibility executive participation in the process.

In his book Values-Driven Safety, Don Eckenfelder emphasizes that one's actions are a moving picture of one's beliefs. In this book and his summary article: "It's the Culture, Stupid" (Occupational Hazards, June 1997), he presents a convincing case that culture predicts results. He contends that an organization's basic beliefs and values (its culture) impacts its decisions which, in turn, define systems and structures which, in turn, determine manager practices, which directly shape employee behaviors and influence work attitudes, and ultimately, all of which determine the performance outcomes (results) an organization achieves. Or as Stephen Covey has said: "Every organization is uniquely designed to exactly produce the results it achieves." Where executive values are weak, downstream organizational roles, relationships, decision-making criteria, trust levels, and management behaviors will compromise safety, and ultimately high losses and costs will be the predictable outcomes.

ORGANIZATIONAL STRATEGY: "Safe by design ... organizational design!"

The fifth safety strategy is that of Organizational Safety a/k/a the safety management process. This strategy is premised upon the belief that safety results will improve if an organization changes those management systems, structures, and processes, which integrate (enable), or isolate (starve) safety in its operations. This strategy addresses the "roles, relationships, procedures, job descriptions and organizational charts" of the company. Tactics typically pursued under this strategy include creating safety policy and procedure, defining roles, responsibilities, and relationships, budgeting processes, goal setting, developing action plans, measuring and creating accountability for results.

Brooks Carder in his work associating safety strategy to TQM principles emphasizes the critical relationship between organizational structure and operational results: "By focusing only on individual behavior," he claims, "the system potentially ignores at least 85% of the factors controlling safety." (Carder) Based on this premise, an organization that effectively designs safety into its systems and processes through organizational design, job descriptions, communications channels, and performance systems will positively impact manager practices, employee behaviors, and ultimately safety results.

PERFORMANCE LEADERSHIP: "Safety follows the leader!"

The sixth safety strategy is that of Performance Leadership, in behavioral literature sometimes referred to as 'performance management' (although I personally see this as more of a leadership issue). This strategy is based on the premise that safety will improve if an organization changes its management methods and consequence delivery practices (from less punitive to more reinforcing). This strategy addresses the inherent deficiencies of hierarchical 'command and control' management practice. It recognizes that how employees perform, (safe or unsafe) is heavily influenced by how managers manage (positive or negative). In order to optimize safe behavior, managers must create an environment that encourages, recognizes, and rewards 'safe' performance. If managers want more safe performance, they must be willing to recognize and reward safe performance more. This means moving from autocratic styles to participative styles, from hierarchical structures to team environments, from manager 'spans of control' to employee empowerment, and from progressive discipline to reinforcing (motivating) practices. Scott Geller sums this up best: with his belief that: "Managers must act employees into thinking differently." (Geller)

BEHAVIORAL STRATEGY: "Safe is how we do business!"

Finally, we arrive at the seventh and perhaps most critical safety excellence strategy that of Organizational Behavior change. This is the 'keystone strategy,' the strategy which locks all others together into a high performance structure, which when subjected to the pressures and stresses of the work environment strengthens rather than fails. This critical strategy is premised on the belief that an organization will improve its safety results by changing organizational behaviors throughout an organization. True behavioral change strategy addresses what 'all people' do in an organization, not just what front-line employees do. This is the ultimate safety excellence strategy as 'what people do' encompasses:

- Education and training - What Human Resource personnel do;

- Statutory compliance - What Legal, and Regulatory Affairs do;

- Safeguarding and process design - What engineers do;

- Values and leadership - What senior executives do;

- Systems and structures – What line and staff managers do;

- Performance Leadership - What supervisors do; and ultimately

- Organizational Performance – What employees do.

Safety excellence is a function of individual and organizational behavior, both of which are a function of the organization's core values …its culture; that force which determines what really is important in the organization, and which causes everyone to pull together to drive safety through the process.

Safety can't be positioned in any one place in an organization; it must be fully integrated within and assumed by all functions. It must build partnerships with all functions, and provide managers with resources and solutions in their assigned areas of responsibility. This is the 'critical success factor' discovered and embraced by the DuPont Corporation in their long journey to World-Class safety leadership. "Safety responsibility (command authority) must always be a line management responsibility. The safety function must always be a support, never a decision-making authority." (Thomen)

As this Architecture for Safety Excellence indicates, there is a formula for safety success - a formula that combines all the critical success strategies, and it is:

SAFETY SUCCESS = CEOu

Where:

C = The Safety **Culture** (values and leadership) of the organization

E = The Safety **Elements** (Education, Enforcement & Engineering).

O =   The Safety **Organization** (safe designed into operations)

u =   YOU! (the ultimate power of success)-The Force of Excellence

For the past 70 years, business has focused almost exclusively on the 'E' (elements) in this equation; the traditional strategies of Engineering, Education and Enforcement. And, for the most part, has mastered these quite well. The Architecture of Safety Excellence proposes that it's now time to work on the 'excellence strategies' of culture, organization, leadership, and organizational behaviors — the true accident sources in an organization. You are the ultimate power of change in your organization. You must become the architect of safety excellence for your organization's future success.

REFERENCES

Andre, Rae and Ward, Peter D. "The 59 Second Employee: How to Stay One Second Ahead of Your One Minute Manager". *Houghton Mifflin*, Boston, MA. 1984.

Carter, Brooks. "Quality Theory and the Measurement of Safety Systems", *Professional Safety.* February 1994.

Alexander. "Factors In Successful Occupational Safety Programs", *Journal of Safety Research.* December 1977.

Cohen, Crites, Thomas R. "Reconsidering the Costs and Benefits of a Formal Safety Program". *Professional Safety.* December 1995.

Deming, W. Edwards. "The New Economics for Industry, Government & Education". Cambridge, MA: MIT Center for Advanced Engineering Studies. 1993.

Eckenfelder, Donald J. "Values-Driven Safety; Re-engineering Loss Prevention Using Value Inspired Resource Optimization". Government Institutes, Inc. Rockville, MD. 1996.

Eckenfelder, Donald J. "It's the Culture Stupid". *Occupational Hazards.* June 1997.

Geber, Beverly. "Because It's Good for You! Bill Clinton's Training Tax". *Training Magazine. Lakewood Publications.* April 1993.

Geller, Scott. Presentation at Alabama Governors Safety & Health Conference, Orange Beach, AL. August 1996.

Johnson, David. Editor. 1998 Readers Poll White Paper. *Industrial Safety & Hygiene News*.

Keck, Paul R. "Why Quality Fails. *Quality Digest*. November 1995.

Manule, Fred. "Safety & Total Quality Principles" Presentation at ASSE Regional Professional Development
Conference, Saratoga Springs, NY. April 1995.

Mr. Quotes – *Business. Red-Letter Press, Inc.* Saddle River, NJ. Copyright 1992.

Paton, Scott M. "Managing Change". *Quality Digest*. November 1998.

"Reality Testing – Assessing the Performance of Workers' Compensation Cost – Management Initiatives".
Tillinghast-Towers Perrin Survey Report. February 1996.

Thomen, James R. "Leadership in Safety Management". *Modern Management Associates*. Wilmington, DE.
John Wiley & Sons, Inc. New York. 1991.

Weaver, D. A. Presentation at Wausau Insurance Safety Management Seminar, Wausau, WI. June 1969.

Safety Excellence Continuum Strategies

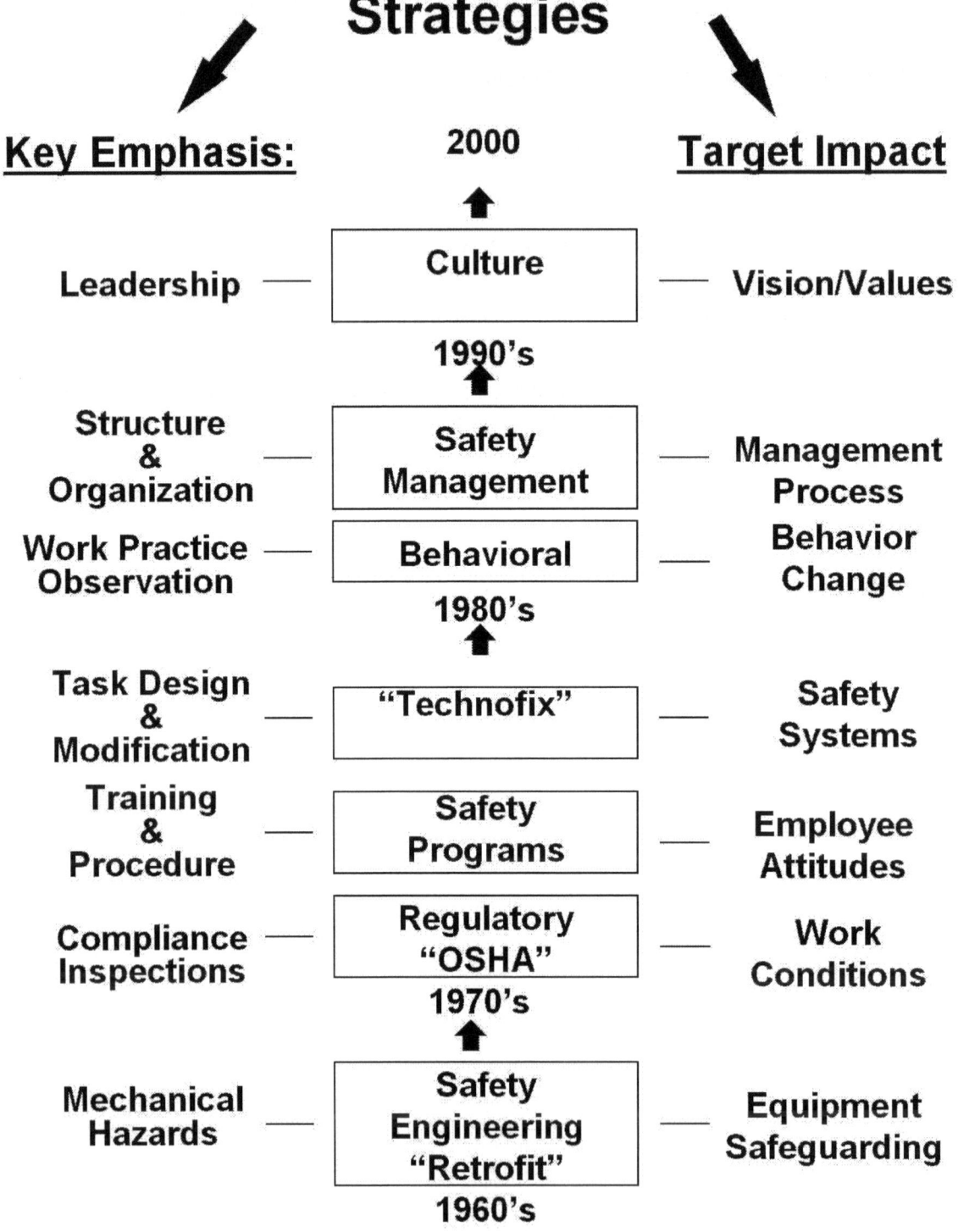

Key Emphasis:	2000	Target Impact
	↑	
Leadership —	Culture	— Vision/Values
	1990's	
	↑	
Structure & Organization —	Safety Management	— Management Process
Work Practice Observation —	Behavioral	— Behavior Change
	1980's	
	↑	
Task Design & Modification —	"Technofix"	— Safety Systems
Training & Procedure —	Safety Programs	— Employee Attitudes
Compliance Inspections —	Regulatory "OSHA"	— Work Conditions
	1970's	
	↑	
Mechanical Hazards —	Safety Engineering "Retrofit"	— Equipment Safeguarding
	1960's	

THE TWELVE PARAMETERS OF MANAGING RISK

1. **Cost** - The total sum of an organization's incurred (insured, retained and/or transferred) risk plus associated fines, penalties and legal expenses.

2. **Loss** - The total value of insured and self funded medical and indemnity claims and the reasonably estimated value of compromised work efficiency and sub-optimized operational performance caused by accidents.

3. **Claims** – Those 'Incurred' (reported and IBNR) injuries, and occupational illnesses arising from operations and causally related to work processes.

4. **Injury/Illness** - Those physically and/or emotionally damaging outcomes resulting from either an accident (acute event) or a disabling (long term) exposure to injurious acts or elements.

5. **Accident** - An unplanned, unintended, and unexpected event…or the planned, reasonably foreseeable, but unintended outcome of practices or processes not effectively managed, controlled, or safeguarded. (IBD- Injuries by design)

6. **Regulatory Compliance** - Those activities and interventions designed to comply with the minimum requirements of statutory rules or regulations.

7. **Safety Programs (Awareness)**. Those programs designed and implemented to build positive and compliant 'safe work attitudes' relying primarily on training, rules, procedures and other behavioral antecedents. i.e. 'the written rules'

8. **Process Design and Safeguarding** - Those mechanical safeguarding initiatives and technical engineering solutions, which proactively design safe, work environments and reactively safeguard identified hazards and accident exposures.

9. **Behaviors** - The actions and in-actions of people throughout an organization, which determine the degree of risk inherent in common, work practices and processes.

10. **Performance Leadership** - Those programs, procedures and practices that reflect the predominant beliefs of managers and common methods employed to shape work performance.

11. **Management & Organization** - The formal, management designed, authority, decision-making and communications structures and systems that define responsibility and information flow within an organization.

12. **Safety Culture** - Those shared assumptions, beliefs, values and visible leadership practices which signal 'what's really important' in an organization; i.e., 'The 'unwritten' rules."

Strategies for:

Managing Organizational Risk and Minimizing Operational Loss

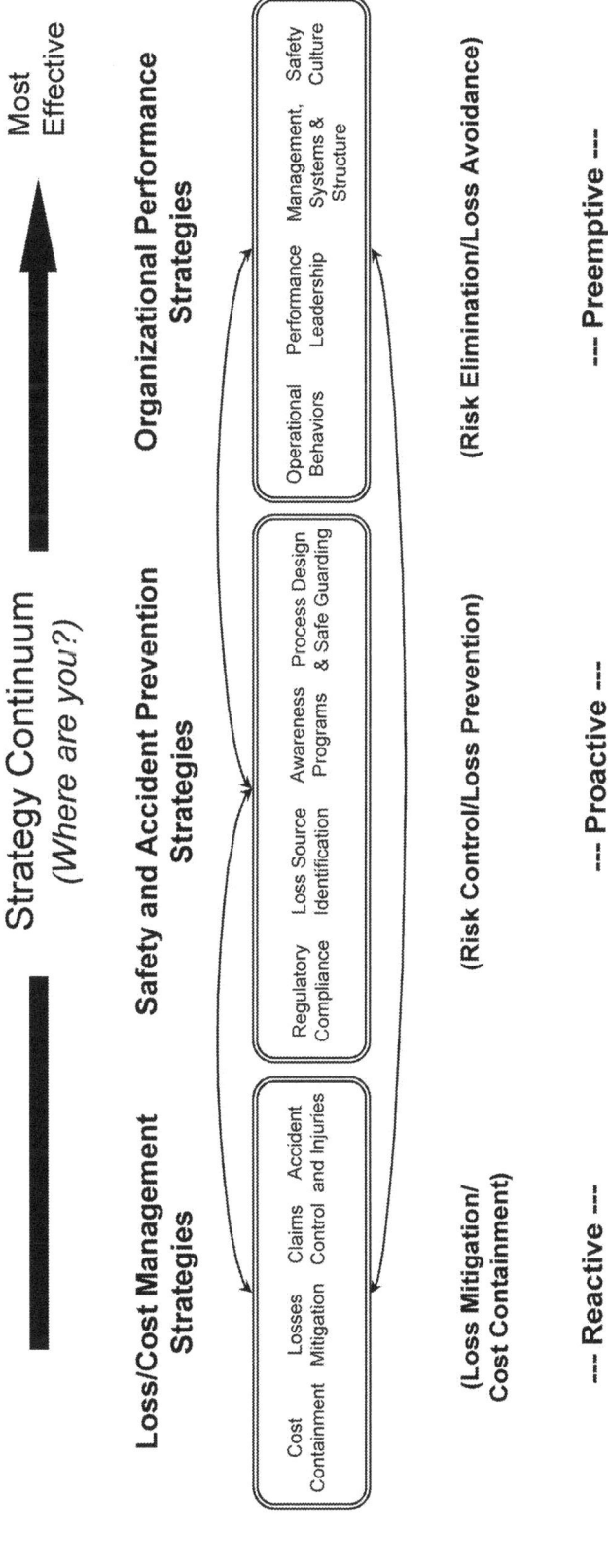

Strategy Continuum
(Where are you?)

Most Effective

Loss/Cost Management Strategies

Cost Containment	Losses Mitigation	Claims Control	Accident and Injuries

Safety and Accident Prevention Strategies

Regulatory Compliance	Loss Source Identification	Awareness Programs	Process Design & Safe Guarding

Organizational Performance Strategies

Operational Behaviors	Performance Leadership	Management, Systems & Structure	Safety Culture

(Loss Mitigation/ Cost Containment)

(Risk Control/Loss Prevention)

(Risk Elimination/Loss Avoidance)

--- Reactive ---

--- Proactive ---

--- Preemptive ---

Strategies which focus on post-injury response to mitigate loss and minimize financial consequences (costs) of accidents.

Strategies which focus on accident prevention via engineering solutions, physical modification, safeguarding, safety training and awareness, and loss source elimination in work processes.

Strategies which focus on organizational effectiveness and which plan, design, and integrate systems and processes which assure safe work practices and accident free outcomes.

Lack of Safety is: **"Financed"**

Safety is: **"Programmed"**

Safety is: **"Integrated"**

'RE-BRAINING' CORPORATE SAFETY AND HEALTH

LARRY L. HANSEN

Originally Published as October 1995 cover/feature of 'Professional Safety' magazine

If the business sages are correct, corporate America has but five short years to complete its transformation order to meet the challenges of the new millennium. Based on progress to date, it appears the hardest challenges are yet to come.

Efforts to re-engineer, automate, computerize, downsize and restructure the corporation have not produced the productivity gains envisioned. In fact, results have been far from stellar. Computerization has not significantly increased white-collar productivity, while downsizing has, in some cases, lowered productivity. One must wonder (as did Peter Drucker): Is doing things 'righter' really the answer?

These efforts, focused on structure, overlook the true source of productivity: PEOPLE. The new frontier for productivity enhancement is 're-braining' the organization – a shift to doing right things right!

Re-braining safety requires a major shift in current beliefs about what drives safety performance. The challenge is set forth here, and 12 guiding principles that will drive the process are discussed.

"We cannot expect someone who works for an insurance company to give advice on the use of their products and services to solve our workers' compensation problems. That's a lot like following free advice on hen house security from the Red Fox Alarm Co."

- Brent Winans

Such is the voiced opinion of one enlightened risk manager and likely the silent perception of many. With losses and insurance costs continuing to escalate as senior managers focus on "re-engineering the corporation', the heat is on to stop the bleeding. This heightened priority has exposed numerous realities; everything is open to review.

QUESTIONING CONVENTION

Risk managers are questioning the conventional wisdom of traditional safety philosophies. Based on the 'hard-number' evidence, these questions are valid. Insurers and insureds alike have misdiagnosed and ignored opportunities to alter lackluster results produced by the current safety paradigm. Business remains loyal to the status quo, content in following prescribed ways and complacent in "doing things right.'

Price Prichett, an organizational change consultant, cautions that in times such as these, "The need for change is most evident [in] the results produced by people doing the wrong things flawlessly." He contends that organizations must "face reality: do what works." With the nation's output of 'human scrap' (workers' compensation) approaching $60 billion, it is clearly time to 'think' again.

Karl Albrecht says: "When one starts with wrong assumptions, applies quick for reasoning, then delegates to a committee, it is hard to arrive at sound conclusions." Yet, this is the typical approach in safety, "Success ultimately depends on the assumptions from which you work, it they are wrong to start with, it doesn't really matter what you do" (Koln).

The message is clear. The future belongs to those willing to question current ways of doing business, abandon existing mindsets, and break the rules that bind business to the past.

SEPARATING MYTH FROM REALITY

When "Safety Management: A Call for (R)evolution" was published (Professional Safety, March 93, pp. 16-21), some recognized its challenge as radical and these people should be applauded. Why? 1) They are 'thinkers', they care enough about the future to participate in it; and 2) They are 'risk takers' (an uncommon breed in today's organizations).

If radical means being willing to ignore 'prevailing wisdom' in favor of seeking the 'uncommon logic,' and if revolutionary means a willingness to 'question traditional ways' rather than 'blindly accept the status quo,' then, without question, I am all of these and more. The truth is: "There are no legitimate theories for success, there are only actions: highly intelligent, not so bright and absolutely stupid. The common theme among all that are highly intelligent is that they work." (Heller).

It is time to separate myth from reality. Safety and health practitioners need to abandon the prevailing (but false) wisdom of tradition in favor of the 'uncommon logic' of success. This necessitates a shift in premise from: People as the problems. "Accidents are the result of unsafe employee acts and behavior." To the 'uncommon logic' that the:

Process is the problem. "Accidents are the result of flawed management values, decision and practices." This represents a difficult change for many, and will be strongly opposed by those heavily vested in tradition. However, as paradigm pioneer Joel Barker proclaims: "Those who say it can't be done need to get out of the way of those who are doing it."

Following are 12 guiding principles that will drive the re-braining of Safety 2000: Success via the Uncommon Logic.

NO. 1: IT'S IN THE ORGANIZATION

Prevailing Wisdom: If a company has escalating WC costs, it definitely needs to implement a safety program.

Uncommon Logic: If a company has escalating WC costs, it most likely has organizational problems that no safety program will fix.

In the late 1970's, Alex Cohen, Robert Cleveland and Harvey Cohen conducted a series of studies to identify practices that equal good and poor safety results. They studied two issues: traditional safety elements and basic management competencies. Results clearly indicated that implementation of traditional safety programs had minimal impact on accident rates.

These organizations were also examined from the standpoint of basic management competencies (i.e., planning, efficiency, budgeting, quality of supervision, communication and employee relations) Based on these criteria, accident rates were differentiated and stratified. The key finding: Companies that effectively managed core business processes produced superior safety results.

These studies reveal a core problem with traditional safety: Safety is perceived only as a 'program' and is typically a 'staff responsibility.' To succeed, safety must belong to line management. American Spring & Wire Corp., Bedford Heights, OH, has recently achieved a significant turnaround in its WC losses. When asked to identify the key drives of this turnaround, Jim McDonald, company vice president, concluded, "I learned that to solve a safety problem, you need to approach it thinking "I'm to blame…no one else" (Avers).

"Success takes more thinking, failure takes more time."

~ Hank Sarkis

"We're running out of time!"

L. L. Hansen

NO. 2: BUILD POSITIVE EMPLOYEE RELATIONS

Prevailing Wisdom: Employee accidents drive WC costs.

Uncommon Logic: Employee claims drive WC costs. People frequently confuse accidents with claims. They are not, in fact, one in the same; one costs large amounts of money!

Traditionally, the insurance industry has built an incurred-but-not-reported (IBNR) factor into loss reserving practices. This charge funds losses that have occurred, but have yet to be 'claimed' because of a delay in either injury manifestation or reporting. Today, a new phenomena exists: reported-but-not incurred (RBNI). It, too, is a charge embedded in WC costs; in this case; however, the employer determines the amount.

Dennis Brooks, president of Comp Management, Inc., Long Beach, CA, believes all businesses need a "claim deterrent process (CDP),' a strategy that goes beyond accidents to address their ultimate outcome claims.

Claims (the dollar value of accidents) are to a large extent subjective, a matter of employee perception and attitude. Employees involved in accidents often sustain injuries that may, or may not, lead to a claim. The decision to file a claim, lose time, extend leave or return to work, and the ultimate degree of residual disability, are choices employees make based on their perception of the organization and its management. "Why do some workers remain on the job while others with similar ailments file for worker' compensation?" asks Presley Reed, psychiatrist and occupational health consultant. "Because," he suggests, "disability is as much a state of mind as it is a state of body."

Human resource practices offer great opportunity to shape attitudes and reduce WC 'claim costs.' When a company fails to build positive employee relationships, it simply fuels the 'claim development process.'

"If you're running your business right, people are going to stop throwing hand grenades."

-Ed Walsh

NO. 3: JUST BECAUSE IT'S TRADITIONAL

Prevailing Wisdom: Traditional safety programs are valid and well founded; they work.

Uncommon Logic: Traditional safety programs are more conventional than wise, frequently lies, they make work.

The 'truth' about safety program effectiveness can be found on Route 281 near San Antonio, TX, where a billboard proclaims (in all capital letters): Texas Country Fried Steak – Voted the Best in the Nation. Printed below (in small letters): Almost 3 dozen sold! The truth is in the numbers…and the nation's numbers do not suggest that safety is winning.

Anthony Veltri, an Oregon State University professor, conducted a survey to determine safety strategies most frequently employed in workplaces. The predominant strategy (77 percent): "Reluctant compliance,' which calls on the safety department to shield the line organization from regulators and ensure statutory compliance. Speaking to industry's focus on results (quality) via compliance, Koln says: "If temporary compliance is the goal of managers, then we just explained the problem with U.S. industry. Temporary is obviously inadequate. As for compliance, quality never comes from mindless obedience." Neither will safety!

"If 50 million people say a foolish thing, it's still a foolish thing."

- B. Russell

NO. 4: MANAGEMENT ACTION REQUIRED

Prevailing Wisdom: Management 'commitment' is the key to overall safety success.

Uncommon Logic: Management 'action' is the sole requisite to achieving overall safety success.

Talk is cheap. Most safety programs are a lot of talk. 'Commitment' is a passive state and can never direct the complex interactions needed to improve an organization's safety performance. Only active involvement can overcome the corporate inertia that inhibits an organization from attaining higher levels of safe performance.

Simonds and Shafai-Sahrai confirmed this via their research of businesses located in Michigan. They identified 11 matched pairs of companies comparable in most demographic categories except for accident outcomes. One set had extremely high accident rates, the other extremely low. Analysis of operational differences between the two demonstrated that companies which "followed through' by acting on their commitment produced safer outcomes.

Commitment without action only produces 'cynicism,' which is typical of employee reaction to 'write 'em and post 'em' corporate policies that proclaim: No job is so important that we (employees) can't take the time to do it safely. Employee response: Why can't they (management) take the time to design it right in the first, place so we don't have to take the time to fix it out here? As Tom Peters says, "They watch your feet not your lips."

However, an interesting paradox arises concerning this requisite for 'executive action.' It is best demonstrated in a large U.S. consumer product corporation, where safety is not dealt with above the mid-level of the organization. This firm believes so strongly in safety as a corporate value that it need not call on its CEO to drive the process. Safety has become an inherent expectation, fully integrated into all processes. Simply put, safe is how things are done, no exceptions; it is not a program.

"It is not enough that top management commits itself, they must know what it is that they are committed to. Action is required."

~W. Edwards Deming

NO. 5: MANAGEMENT CREATES BAD ATTITUDES

Prevailing Wisdom: Poor employee attitudes cause the WC problem.

Uncommon Logic: Poor management practices cause employee attitudes, it is not a matter of fate.

All business issues ultimately are reduced to 'make or buy decisions.' Without question, employee attitudes (poor as they may be) are a 'make' decision by managers.

Business has invested heavily in 'selecting out' problems by dedicating an entire corporate function (personnel) to design and implement procedures to 'select in' right people. Such efforts have been successful. Managers do not intentionally hire 'bad attitudes.' This leaves but one conclusion. If bad attitudes are prevalent, managers are highly efficient at 'making them.'

Bad attitudes are an issue but not the problem. Their cause, the reasons bad attitudes exist – specifically, the practices that create them – is the problem. In The Customer Comes Second, Hal Rosenbluth says, "business earns the bad attitudes of its employees."

Employee attitudes are a 'reaction' to management 'actions' (a make decision). Attitudes span a spectrum from B.A.D. (Belligerent And Destructive) through average J.O.E. (Just Ordinary Employees) to S.A.I.N.T., those to (Say All Injuries are Negligible and Temporary). Employees position themselves along this spectrum based on how they are treated. In other words, some companies take advantage of their employees (maximizing them as a resource), while others take advantage (disregard or exploit) of their workers. Employees react accordingly!

The "Law of Subordinate Superpower" is a peculiar phenomenon in employee relations. Unlike the laws of physics, which state that "for every action, there is an equal and opposite reaction," this phenomena holds that "for every manipulative management action, there is an employee reaction, which will definitely be opposite but will never be equal!" Michael Shor, president of Health Care First Inc., agrees: "The best loss control program in the world can never make up for [an organization's] lousy employee relations."

"Employee attitudes are important, but the fact is they are irrelevant until management attitudes are addressed."

~ J. Michael Crouch

NO. 6: IT'S THE PROCESS, NOT THE EMPLOYEES

Prevailing Wisdom: Unsafe employee acts are responsible for 85 percent of all accidents. In other words, employees are the problem.

Uncommon Logic: The process, designed and administered by management, is responsible for 94 percent of all outcomes (including accidents). In other words, management makes the majority of the mistakes!

Tom Peters speaks of "a blinding flash of the obvious," a phenomena in which obvious facts simply do not lead to obvious conclusions. Such phenomena definitely exist in safety. Managers typically say that the production process (planning, organizing, staffing, developing specs, budgeting, specifying materials, establishing rules, designing layouts, etc.) is a management responsibility.

Yet, when employees are injured thanks to this process, what is management's typical reaction? "Careless employees!" Wrong! Employees sustain injuries, and accidents occur, due to the process, which is designed, operated and owned by management.

A large Midwestern retailer plagued with high WC costs issued a corporate directive identifying the 'real causes' of accidents within the organization:

- Employee lack of respect

- Employees being 'above rules.'

- Employee retaliation

- Employee incompetence

- Employee indifference

Contrast these pronouncements with the values of Proctor & Gamble Corp. At P&G, employees are:

- Essential to the ongoing success of the enterprise

- Entitled to preservation of health

- The key to productive, high-performance work systems

Commenting on the value of employees, P&G's CEO said, "[We] could lose all [our] plants in a single major catastrophe and conceivably be back in business with restored market share within 10 years. If we were to lose our people, there is no return…there is no future…it's all over" (Fulweiler).

> *"The American work ethic is alive and well, urgently wishes to express itself, and is hobbled at every turn by management."*
>
> ~ Daniel Yankelovitch

NO. 7: THINKING IS CRITICAL

Prevailing Wisdom: Compliance with safety rules ensures safe operations. 'Obedience' is, therefore, required.

Uncommon Logic: Rules can never adequately address hazard variables inherent in a dynamic organization. Thus, 'thinking' is critical.

Obedience and thinking are at opposite ends of the business spectrum, directly aligned with failure and success. Progressive companies recognize that success is not achieved via 'rules'; employees will follow rules (no matter how ridiculous). Dana Corp. attributes much of its success to the fact that the company 'burned its procedure manuals.' Dana Corp. understands that rules promote blind compliance, while real success is driven by 'thinking.'

My son, Eric, a third-year business student, experienced 'real-world' compliance management during a summer job. He calls the experience 'anti-think/double think,' which he describes this way: "In the past, business was operated under the premise that managers did the thinking and employees did the 'doing' (i.e., no thinking allowed)." New philosophies call for empowerment, participation and employee involvement, which, he observed, is really just a ploy (anti-think/double think). Managers say they want employees to participate and offer opinions; yet, when employees do become involved and tell managers what is really wrong, employees are ignored, chastised or labeled 'not team players.' America's workplaces do, indeed, need to be 're-engineered;' what is needed is more employee 'head room.'

*"Regulations are for the obedience of fools
and for the guidance of wise men."*

~ R.A.F. Motto

NO. 8: EFFORT NEEDED UPFRONT

Prevailing Wisdom: Safety inspections are a timely, effective way to identify problems and prevent serious accidents.

Uncommon Logic: Safety inspections rarely identify real causes of accidents and only defer, in time, their ultimate occurrence.

Abraham Lincoln once said: "If I had eight hours to cut down a tree, I'd spend six hours sharpening the ax." Success requires time and effort upfront – planning, organizing and facilitating a process, rather than at the end, correcting mistakes. Traditional safety programs devote little time to critical upfront issues, consequently, most time is spent after-the–fact, patching holes.

True accident causes rarely lie on the production floor; symptoms do. Real causes are found in corporate offices and planning rooms, places not frequented by safety directors. Managers should cease reliance on inspecting hazards out of the process and dedicate efforts to designing safety in.

*"Man will occasionally stumble over the truth, but most of
the time he will pick himself up and continue on."*

~ Winston Churchill

NO. 9: FIND THE REAL ACCIDENT CAUSES

Prevailing Wisdom: Accident investigations reveal critical facts that prevent accidents from recurring.

Uncommon Logic: Accident investigations rarely identify real accident causes, which are embedded deep within the organization. Therefore, recurrence is clearly inevitable.

If one believes the findings of most accident investigations, then the real causes of workplace accidents are:

1.   Careless employees: 40 percent

2.   Beats me. Damn…I dunno! 25 percent

3.   All other: 35 percent. This 'catch-all' category would include: employee carelessly used broken ladder, employee, without thinking, plugged defective tool into power source; inattentive employee fell over crate in the aisle; employee was performing normal job and back started to hurt; or distracted employee became trapped in unguarded machine.

Obviously, such conclusions are open to question yet seldom are! The problem: Accident investigations are a responsibility placed at a level within the organization (first-line) that cannot truly address real accident causes – upper level management decisions. If accident investigations do not identify system failures, they do not produce accurate information. Most do not!

"Everybody lies…but it doesn't matter; nobody's listening."

Lieberman's Law

NO. 10: NO QUICK-FIX SOLUTIONS HERE

Prevailing Wisdom: Safety incentive programs are quick, easy and inexpensive; they drive safety improvement.

Uncommon Logic: Safety incentive programs are quick, easy and inexpensive – sufficient evidence that they do not work!

Here is a tip: Buy stock in safety incentive corporations; they are a growth industry. As the WC crisis deepens and executives become aware of the real costs of accidents, they will frantically seek quick, easy solutions: Viola – incentive programs!

The truth remains, however, that over time, incentives do not produce lasting results. In Punished by Rewards, Koln identifies key reasons why these programs have little impact on long-term accident costs:

•   They are only incentive programs. They do not obligate change in existing processes or procedures.

•   Incentives ignore reasons. They disguise genuine deficiencies and strategic flaws that exist within the organization and/or process.

Add one more reason to this list: They are premised on 'wrong-headed' assumptions that accidents are intentional acts and that a baseball cap, belt buckle or savings bond will cause employees to stop placing limbs into unguarded machines.

One upstate New York manufacturer attempted to address high accident rates and poor employee relations (typical companions) via a monthly drawing for baseball tickets. The industrial relations manager, excited about the first drawing, planned a ceremony with free coffee in order to visibly demonstrate management interest. No one showed up! Employee involvement cannot be won by a few game show prizes.

During a recent interview with Safe Workplace, Peg Seminerio, Director of Health and Safety for AFL-CIO, explained, "The use of incentives and rewards sends the wrong message to workers. There's growing concern that incentive programs don't necessarily [address] underlying problems in safety." Koln adds: "There is an important thought process regarding the question, do incentive plans work? Many are willing to say not now. Not so many are willing to say not ever!"

> *"There's always an easy solution to every human problem, neat, plausible...and wrong."*
>
> ~ H.L. Menken

NO. 11: YOU CAN'T TRAIN THE PROCESS

Prevailing Wisdom: To improve safety, an organization must make a significant commitment to employee training.

Uncommon Logic: To improve safety, an organization must make a significant commitment to fix whatever is truly wrong (which is generally not employees).

Organizational problems – deficient planning, poor organization, unclear goals, lack of vision, vague responsibilities, autocratic direction, lack of employee involvement, conflicting priorities, poor communication and incompetent supervision – are the real accident sources. When these factors interact and culminate in accidents, management's frequent response: "We need a training program." Such a reaction says 'people at fault' rather than 'process at fault.'

As W. Edwards Deming noted, management is responsible for most outcomes of the production system, including its volume of 'human scrap.' The process needs fixing 94 percent of the time...not the people.

"No amount of care or skill in workmanship can overcome fundamental faults of the system."

~ W. Edwards Deming

NO. 12: SAFETY SITS ON THE BOARD

Prevailing Wisdom: Safety is an employee issue that is most effectively handled by the personnel department and safety committees.

Uncommon Logic: Safety is a boardroom issue, which can only be impacted by that group.

Accident costs are no longer a negligible pass-on expense that can be ignored or buried in the cost of doing business. Escalating WC costs have truly become a boardroom issue. In the auto industry, for example, employee accident and health costs are now a major raw material cost of manufacturing a vehicle. In other industries, these costs often exceed 50 to 75 percent of payroll.

Yet, how do corporations typically deal with such problems? By creating staff/employee committees that lack direction, time, funding, and authority needed to truly impact real (organizational) accident causes. The result: Monthly meetings (whether needed or not) and the predictable 'bitch list.'

Five years ago, Hoechst Celanese Corp. changed its approach to safety. At that time, says Dave Johnson, safety manager, the company maintained industry-average incident rates, and its safety program was 'traditional.'

Reality hit when a severe accident and audit report from a major customer negatively impacted revenue. This turn of events impacted decision-makers and prompted a strategic rethinking of safety as a core value. Incident rates have improved each year since that organization's transformation.

Results such as these cannot be produced through monthly safety committee meetings. Such results can be produced by a 'board of directors' and they usually only meet quarterly!

"Lots of people confuse bad management with destiny."

~ E.Hubbard

REFERENCES

Albrecht, Karl. The Only Thing That Matters. New York: Harper Collins Publishing, 1992.

Austin, Nancy and Tom Peters. A Passion for Excellence, New York: Random House Publishing, 1985.

Avers, Laura. "Strengthening Safety: An Evolutionary Process." *Ohio Monitor.* March/April 1994.

Carder, Brooks. "Quality Theory and the Measurement of Safety Systems." *Professional Safety.* Feb. 1994: 23-28.

Cohen, A. "Factors in Successful Occupational Safety Programs." *Journal of Safety Research.* December 1977.

Crosby, Phillip. Quality Is Free. New York: *Penguin Group,* 1980.

Deming, W. Edwards. The New Economics. Cambridge, MA: MIT Center for Advanced Engineering Studies, 1993.

Fromm, Bill. The 10 Commandments of Business and How to Break Them. New York: G*P Putnam's Sons,* 1992.

Fulweiler, Rick. Presentation at Minerva International Conference, Scottsdale, AZ, May 1994.

Heller, Robert. The Naked Manager. New York: *Truman Talley Books,* 1985.

Johnson, Dave. Presentation at Minerva International Conference, Scottsdale, AZ, May 1994.

Koln, Alfie. Punished by Rewards..New York: Houghton Mifflin Co., 1993.

Minter, Stephen. Editorial. *Occupational Hazards.* July 1994.

"Organized Laborers Viewpoint." *Safe Workplace.* Spring 1994.

"Safety Program Practices in Record Holding Plants." Morgantown, WV: *U.S.Dept. of Health Educational Welfare*, NIOSH, Div. of Safety Research, March 1979.

Sarkis, Hank. "What Really Causes Accidents." Research and presentation at Wausau Insurance Safety Excellence Seminar, Canandaigua, NY, June 1990.

DELIVERING A 'ONE-TWO COMBINATION' TO FLATTEN L.O.S.S. COSTS!

LARRY L. HANSEN

Originally Published as October 2003 feature in' Occupational Hazards' magazine.

*"O' Lord grant me but one more hard cycle...
and I'll promise not to screw it up!"*

-Insightful Business Executive (Seat 2B)

Hold-on...Here we go again! The 1999 predictions by A, M. Best and Conning & Company concerning workers' compensation costs in the new millennium have come in 'right on the money'. "While there is uncertainty about where rates (workers' compensation) are headed, the outlook on losses is much clearer – they are heading up!" (MaryAnn Godbout, OH&S, August 1999). And, as predictable as are the laws of physics, when losses increase (and investment income plummets), insurance premiums have but only one direction to go... UP!

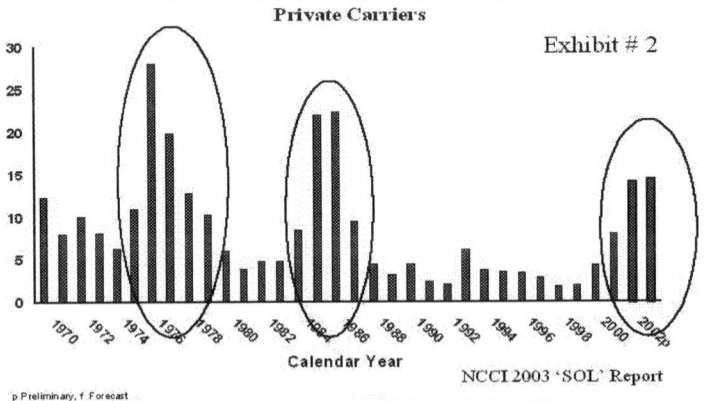

If it seems like we've been here before, you are right! According to NCCI historical data we've been at this point in the cycle, three times over the past three decades. The past soft market of intense price competition (a/k/a creative underwriting) has slipped away (more like vanished overnight), and we're now faced with the challenge of solving 'real' loss problems. To do this successfully, we need focus on the two critical issues that most impact costs: better management of what causes accidents (process), and better leadership of what causes claims (people)! What's needed is a 'one-two combination' to flatten workers' compensation costs!

On a PBS radio program discussing the state of business and economy, I mused as the commentator described the challenges that both 'for profit' and 'nonprofit' organizations would be facing in the future. He made this insightful observation: "About 30 percent of our nation's businesses are nonprofits, but only half that number are actually chartered to be so!" The program further described how a grant application from a nonprofit organization was recently rejected by the funding authority because: "the financials looked too much like a 'for profit' corporation."

These observations make it clear that charter differences, in reality, have been erased. Being in business today, is being a business today. The distinction between 'for profit' and 'nonprofit', in practice, have faded. The bottom line is -- the bottom line! Or as Steven Covey puts it: "If there is no margin, there is no mission."

Profit orientation, however, is not the key to an organization's long term financial health and sustainability; the reason 'why' is. It's a businesses' leadership values and management process, which leads an organization to long term success...or failure. For all businesses, irrespective of charter (profit, nonprofit, or profits that aren't), success is not a top line (revenue) or bottom line (margin) issue...success lies in the ability to effectively manage 'the middle lines' (L.O.S.S., cost and expense. According to Peter Drucker: "The first duty of an organization is to survive and the guiding principle of business economics is NOT the maximization of profit, it is the avoidance of loss."

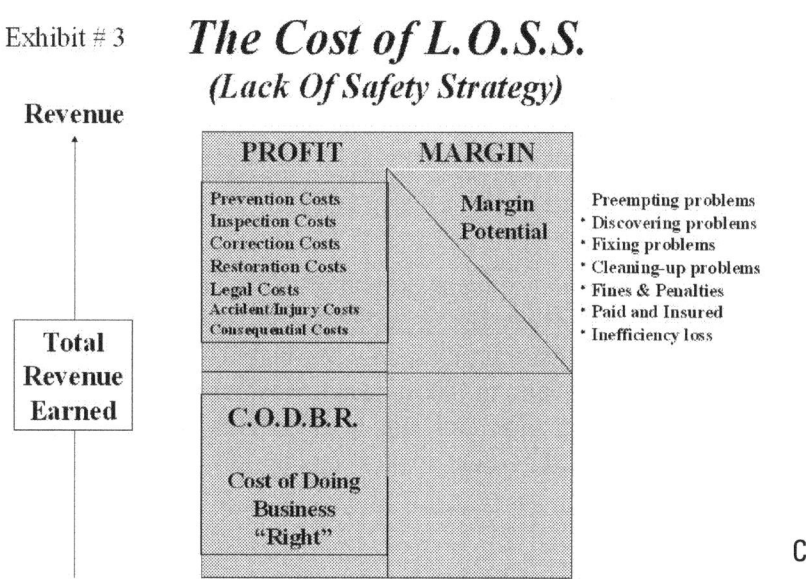

Exhibit # 3 — *The Cost of L.O.S.S.* (*Lack Of Safety Strategy*)

Credit to Ron Glasgow

One of the most over looked opportunities for margin contribution is the management of operational 'L.O.S.S'. – Lack Of Safety Strategy. Unfortunately, most companies fail to effectively measure L.O.S.S. in their organizations, hence have little perception of the financial burden they carry, or appreciation of the margin opportunities being missed. Don Eckenfelder, in his book: "Values Driven Safety', illustrates the impact that effectively managing L.O.S.S. can have on the bottom line of an organization, a range he suggests that can span from substantial for most companies…to perhaps the single largest margin contributor for others. For a billion dollar corporation, he provides this compelling example of the untapped value available from effective L.O.S.S. management.

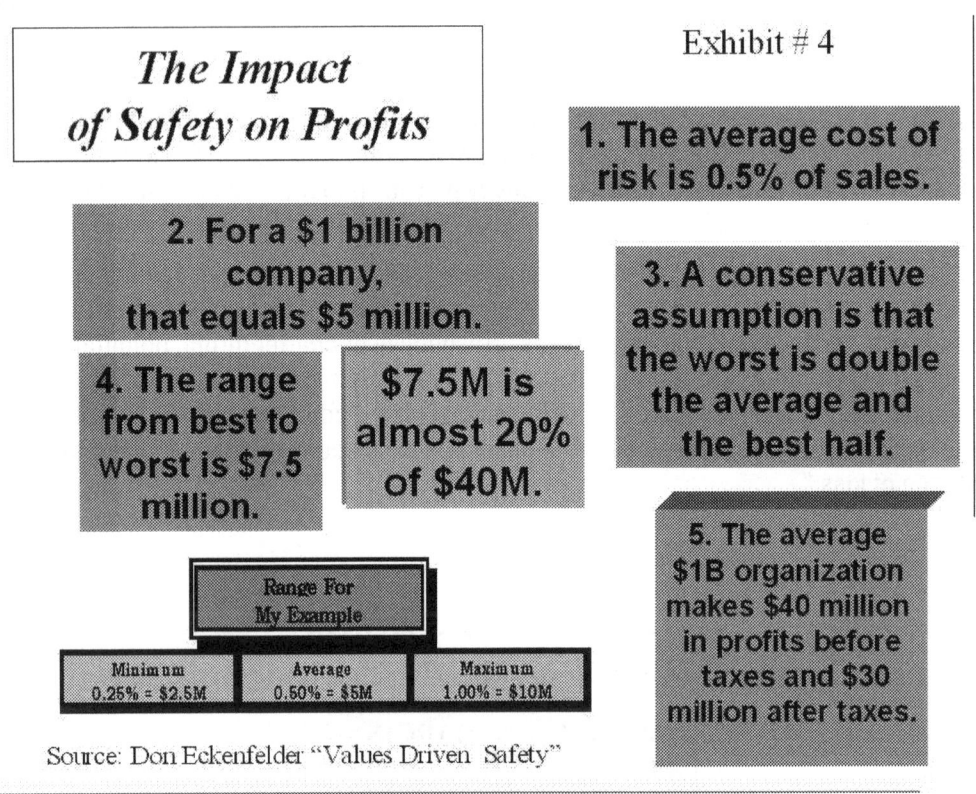

The Impact of Safety on Profits

Exhibit # 4

1. The average cost of risk is 0.5% of sales.

2. For a $1 billion company, that equals $5 million.

3. A conservative assumption is that the worst is double the average and the best half.

4. The range from best to worst is $7.5 million.

$7.5M is almost 20% of $40M.

Range For My Example

| Minimum 0.25% = $2.5M | Average 0.50% = $5M | Maximum 1.00% = $10M |

5. The average $1B organization makes $40 million in profits before taxes and $30 million after taxes.

Source: Don Eckenfelder "Values Driven Safety"

The need to effectively manage L.O.S.S. is the common ground of all businesses, irrespective of profit orientation. L.O.S.S. control is critical to success. Unfortunately, evidence suggests that managers are poorly schooled in the financial sciences. A business literacy survey conducted by the Association for Quality and Participation polled 120 randomly selected Fortune 1000 decision-makers, and found that only nine percent could answer 10 basic business and finance questions. And more enlightening, 63 percent believed that a company could operate without cash, and 34 percent couldn't identify key information on a business income statement." (Usilander)

Some might question these findings, but quite frankly, I don't! In my work as an organizational performance consultant (with a focus on risk management), I've found that the biggest impediment to safety success is executive ignorance! (Okay, now that you've retrieved this from the trash can… just out of curiosity, of course, read on.) By ignorance, I do not mean that executives are not brilliant individuals; they most definitely are! What I do mean, is that most Senior Operations managers have little understanding of how insurance costs are derived, how accidents impact costs, or how organizational initiatives can effectively minimize accidents, injuries, claims, and ultimately the high insurance costs they generate. But, it's not their fault. Business schools don't deal to any large extent with risk management issues, and if they're not otherwise informed, how could they know? To counter this state of misinformation, I frequently engage line managers in a session entitled: "Insurance Premium, Magic Bucks, and Other Mystical Notions". It's a session on Insurance Costs 101, and is designed to clearly 'connect the dots'-- losses to costs! When provided with accurate information, executives have no problem seeing opportunities, strategizing solutions, and producing high levels of success. They just need the 'one - two combination' of facts. The first (jab) being that accidents are due to process deficiencies, a function of management, and the second (right cross), being that operational loss (injuries and claims) are 'people issues', a function of leadership. And, most importantly that both are management responsibilities and opportunities!

Successful organizations recognize that average never is good enough. In support of that position is this interesting insight. A major workers' compensaiton insurer attempted to profile group desirability norms by analyzing the distribution of Experience Modification Ratings (EMRs) within one of its insured associations consisting of over 270 member companies. The findings were insightful. Over two-thirds of the group (181 members) had modifications between .60 and .99 (better than average). One-third (88 members), had modifications between 1.01 and 1.79 (worse than average). And, only one organization had a modification of 1.00 – only one was average! This tells us, as business managers, we really have only two choices: to succeed or fail; to be better than average or worse – there really is no middle ground for high performance.

Assuming that success is the goal, what must a risk manager do to flatten workers' compensation costs? The short answer: Demand answers to these critical questions…Inquiring minds want to know:

1.   WHY'…do we spend so much money on 'safety programs' and still have so many accidents?

2.   WHY'…do we invest so much in employee training and still have unsafe behaviors in our operations?, and

3.   WHY'…in spite of accident rate reductions, do our workers' compensation costs continue to escalate?
      —Inquiring minds 'really want to know!

Good risk management isn't having all the answers; it's asking the right questions. Many risk managers, unfortunately, aren't asking these questions of their in-house staff, insurers or loss control service providers. They continue to accept traditional answers and approaches: i.e., invest more (time and money) on the 'S.O.S.S.' that hasn't worked before--more programs, more meetings, more rules, more committees, more inspections, more training, retraining, remedial training, and ultimately more (whack-a-mole) employee discipline!

Unfortunately, more of the 'Same Old Safety Stuff' doesn't produce different (better) results. In safety, things don't add up; you reach the truth by subtracting down. And, when we peel away the myths of traditional safety 'Wiz-dumb', (wrong headed thinking that impedes progress in the right direction), we find that safe isn't an outcome of regulatory compliance, mechanical retro-fixes, or programs designed to heighten employee awareness and change employee attitudes. We discover an inverse relationship to exist—do more traditional safety; get less operational results!

What research has made amply clear is that accidents (alone), don't drive workers' compensation costs – claims do! If we don't effectively manage both the causes of accidents (process), and the causes of claims (people), all we get are lower incident rates and higher loss costs. Major change will occur only when the workers' compensation crisis is acknowledged to be a symptom of something wrong in the management system. There is an unspoken, yet widely held belief that an organization would have few, if any, problems if only workers would do their jobs correctly. In fact, the potential to eliminate mistakes and errors lies mostly in improving the system through which work is done—not changing the workers." (Sukay)

Starting in the late 1970s, studies have been conducted addressing the critical elements of safety success. Of the more notable are those conducted by the National Institute of Occupational Safety and Health (NIOSH) and more recently by the US Department of Energy (DoE).

The NIOSH study evaluated traditional safety elements (i.e., safety committees, safety rules, accident investigations, etc.) and their impact on accident rates. This study revealed that there were no significant differences in the accident rates between organizations that did these things efficiently vs. those that did not. Researchers then revisited the studied population and analyzed accident outcomes compared to effectiveness in core management competencies, (planning, budgeting, attitude toward workers, management/employee relationships, etc.). What evolved was a direct correlation between these 'organizational elements' and accident rates, -- i.e., those companies that managed poorly had high losses, and those that managed well had low losses. This landmark study confirmed that culture and management process more than a safety program determines the level of safe outcome in an organization.

A more recent study by the Department of Energy at selected sites also confirmed a 'process – result' causal relationship. This DoE study examined accident costs incurred at selected sites as compared to the safety budget of those sites. This study reached two conclusions (Crites):

1. "Increased investment in a formal safety program did not produce improved safety performance." In fact, distribution of results indicated an inverse relationship i.e., the greater the safety investment, the higher the level of loss, and

2.   "Factors having minimal impact were:

- A shift in safety emphasis;

- Size of safety budget;

- Degree of hazard;

- Safety rules (quantity or quality), and

- Safety committees."

I doubt that either organization set forth in their research expecting to find inverse results – but they did! Both confirmed that effective loss reduction is more a function of organizational culture and management process than of safety programs.

WHAT REALLY MINIMIZES LOSS COST?

Insurance industry data over the past decade has clearly identified injury management strategies as the 'low- hanging fruit' for workers' compensation loss reduction. In the 1990s, a major brokerage firm surveyed 700 organizations employing 3.8 million people, and generating $1.1 billion in workers' compensaiton losses. This survey determined that the total cost per claim rose 35 percent during the period 1989 to 1991, while the average claim frequency (claims per 100 employees) dropped nearly 10 percent (10.9 to 10.1). This study confirmed that managing the cost of claims (the human element) is equally, if not more important, than managing the cause of accidents (work environment). And National Council of Compensation Insurers (NCCI) data tracked over time further confirms this fact. The NCCI 2003 'State of the Line' (workers' compensaiton) report, shows that incident rates have steadily declined, while workers' compensaiton losses (medical, indemnity, and total incurred) continue to rise.

Although, there are no 'quick fixes' in business, safety, or workers' compensation, findings of a national claim administrator suggest that there are opportunities for rapid returns when an organization invests in aggressive injury management, de-lawyering, and return-to-work strategies.

A national Third Party Administrator (TPA) studied indemnity claim data from 1409 employers with return-to-work programs (RTW) vs. 8,592 employers without, and found:

- Average claim costs for companies employing RTW strategies to be $7,217 compared to $17,944 for employers without.

- Length of disability for claims of RTW organizations to be 2.8 months compared to 5.3 months for those without.

- Average time to Maximum Medical Improvement (MMI) to be 5.1 months for those with RTW compared to 13 months for those without.

- Average indemnity (lost wage payments) for RTW companies to be $6,120 compared to $15,720 for those without, and,

- Average medical costs for RTW companies to be $1,097 compared to $2,224 for those not employing RTW.

Allowing first-line supervisors to make decisions which impede employees from returning to work "unless 100 percent recovered" are costly …often six-figure decisions! It may be time to put a limit on their spending authority!

This message on workers' compensation cost control started with a focus on profitability-- the issues of top line, bottom line, and the critical 'middle lines.' I appropriately choose to end there as well. If an organization wants to 'KO' workers' compensation costs, it must confront the two key drivers of them…management of process and leadership of people. "To be successful at reducing injuries and workers' compensation costs, American businesses must address the quality of management and management systems." (Sukay) Past research, current statistics, and our on-going experience tells us that much of what we've done in the past hasn't worked. To succeed, we must turn top dollars into bottom dollars by managing 'L.O.S.S.' dollars. We will achieve this, not by investing more in traditional safety programs, but by greater efforts to integrate safety into core business processes and organizational value systems. Losses will only be significantly reduced when SAFE becomes how business is done, not a program, and when managers recognize that injury costs are a function of employee attitudes and behaviors, both of which are shaped by leadership practices.

To reap rapid returns in a workers' compensation cost reduction initiative, an organization must employ positive employee relations, compassionate (and financially sound) injury management practices, and aggressive return-to-work strategies. These initiatives will effectively 'de-lawyer' the workers' compensation process. It must then move from traditional safety programs to progressive organizational performance strategies. These two changes, when timed perfectly, will deliver the 'one-two combination' that will successfully 'flatten' L.O.S.S. costs.

"We've rope-a-doped long enough -- it's time to deliver a one-two combination to flatten workers' compensaiton L.O.S.S. costs!"

~ L.L Hansen

PRINCIPLES OF L.O.S.S. MANAGEMENT
(Lack Of Safety Strategy)

1. "To get employees to pay more attention to safe practice, assign more of a manager's pay to effectively managing safe process."

2. "A dollar saved (LOSS) is more than a dollar earned (revenue) when valued at margin."

3. "Managing 'risk' is wise; doling out 'LOSS' is expensive." Spend money on prevention . . . not insurance. Financing LOSS into the future increases its ultimate cost."

4. "Losses allocated to line managers minimizes losses generated in line operations".

5. "If we took greater action on true accident causes , we would have far fewer under-performing managers."

6. "Supervisory accident investigation reports rarely link root cause with effect, and thus are counter-productive, -- would you blow-in your boss?"

7. "If you think hiring a safety director will solve LOSS problems; think again!"

8. "Taking sides in the 'traditional' (conditions) vs. 'behavioral' (practices) safety debate is avoiding excellence, and siding with mediocrity."

9. "How managers manage directly impacts how employees perform…including Safe vs. Unsafe. Always seek the 'good reasons' for poor performance."

10. "Safety programs impact a number of things; accident costs aren't one of them."

11. "Accidents are 'triggered' on the shop floor; accidents are 'caused' in the corporate offices. Managers work 'on the system'; employees work 'in the system'. The system causes accidents; employees sustain injuries—fix the system!"

12. "It defies business logic to believe that so many smart managers can hire so many dumb employees. Someone's got to smarten up!"

REFERENCES

Cohen, Alexander, "Factors in Successful Occupational Safety Programs", *Journal of Safety Research*, December 1977.

Cohen, Harvey H. and Cleveland, Robert J., "Safety Program Practices in Record-holding Companies", *Professional Safety*, March 1983.

Crites, Thomas R., "Reconsidering the Costs and Benefits of a Formal Safety Program", *Professional Safety*, December 1995.

Drucker, Peter F., Technology Management & Society. New York; Harper Row Publishers, 1972

Eckenfelder, Donald J. "Values Driven Safety", Government Institutes, Inc., Rockville, MD, 1996.

Hefrey, Patricia F. "Evaluating the Practicality of Return-To-Work Programs: Increasing Costs and Legislative Changes Make Disability-Related Problems Too Costly to Ignore", Crawford and Company, Atlanta, Georgia, August 1993

Mealey, Dennis, National Council of Compensation Insurers, '2003 State of the Line' (Workers' Compensation) presentation, NCCI Web site, May 2003.

Planek, Thomas W., and Fearn Kevin T, "Reevaluating Occupational Safety Priorities: 1967 to 1992, *Professional Safety*, October 1993.

Salazar, Noe, "Applying the Deming Philosophy to the Safety System", *Professional Safety*, December 1989.

Sukay, Lawrence D., "Safety Programs Alone Don't Work in Reducing Workers' compensaiton Costs", *Risk Management*, September 1993.

Usilaner, Brian, et al, "What's The Bottom Line Pay Back for TQM?", *Journal for Quality and Participation*, March 1992.

Wolf, Harvey J. and Pearson, John C., "Happy Workers Mean Fewer Injuries", *Safety and Health*, 1992.

WHAT'S YOUR ORGANIZATION'S LOSS MANAGEMENT IQ?

LARRY L. HANSEN

Originally Published as November 1999 feature in' Occupational Hazards' magazine.

With early signs of a hardening in the workers' compensation market, it's critical that safety professionals rededicate their organizations to higher levels of loss control effectiveness. The extended soft insurance market, fueled by excess capital capacity and intense market share competition, has diverted management attention from the underlying drivers of workers' compensation costs — accidents and their organizational causes! In the minds of many managers, all is well; they labor under a false sense of financial security.

When workers' compensation costs ultimately re-emerge as a threat to corporate profitability, organizations will need to refocus efforts on effective loss management strategies. Such strategies, however, will require more than an emphasis on traditional safety programs and regulatory compliance which, historically, have demonstrated limited ability to impact incident rates and workers' compensation costs. Studies by Ken Mitchell and Steve Leclair of National Rehabilitation Planners, Inc., have confirmed: "Safety programs, although necessary, are not sufficient to control disability costs." Effective loss management requires managerial attention to both proactive accident prevention and responsive loss mitigation strategies.

An organization's loss management capability must extend beyond its safety and risk management functions and become fully integrated into the line management structure of the organization. Managers must become proficient in the core competencies and critical lessons of loss management success...the following competencies and lessons:

1. Workers' compensation (how to play the game by the rules and win in spite of them!). Lesson: It's okay to rewrite the rules!

2. Safety Strategy (how to recognize and remedy the real causes of accidents in an organization). Lesson: Poor performance has good reasons.

3. Cost Containment (how to minimize that portion of workers' compensation loss not medically imposed on the organization). Lesson: We have met the enemy and it is us!

4. Financial Implications - i.e., How to measure the true financial impact of "L.O.S.S." (Lack of Safety Strategy) on the organization. Lesson: Pay me now...or pay me a lot more later!

These four areas comprise the critical body of knowledge necessary for effective loss management. They represent an organization's 'Loss Management IQ.'

As Will Rogers once said, 'It ain't what you don't know that's a problem, it's what you know that ain't so!" Unfortunately, in many organizations, operation's managers know a lot that ain't so! What's needed is some straight talk about loss management's critical elements: workers' compensation, safety strategy, cost containment, and financial impact on results.

The following Loss Management IQ Test is designed to appraise these issues and reveal organiza

tional misconceptions and/or performance barriers which may be impeding future success.

Since an organization's true ability to control loss lies beyond the safety practitioner, it is suggested that this test be completed by key executives, line managers, administrative staffers and front line supervisors. When tests have been completed, compare the predominant beliefs and common mind sets to the answers and discussion presented in next month's issue of Professional Safety. Exploring the differences in beliefs and/or variance in understandings between functions, shifts, facilities or levels of management in an organization can identify performance impediments and target opportunities for performance improvement.

WHAT'S YOUR LOSS MANAGEMENT IQ?

Originally Published as November 1999 feature in' Occupational Hazards' magazine.

THE TEST

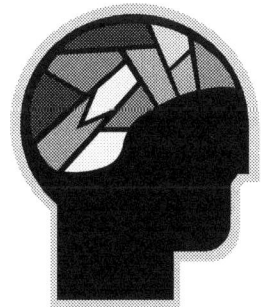

This exercise is designed to address key issues impacting safety performance and workers' compensation costs in your organization. Answer the following questions to the best of your ability, based on your current understanding and beliefs concerning these issues. There are no trick questions in this exercise; it is intended to be an objective, factual and enjoyable learning experience. (P.S.: Watch the person to your left—they're trying to look over your shoulder!)

1. The most expensive (and generally least effective) risk management strategy for an organization to employ in dealing with its workers' compensation costs is _____ _____.

2. **True** or **False**: A 20% dividend on your workers' compensation insurance policy means your company has earned a highly favorable rate of return on its insurance program.

3. **True** or **False**: Workers' compensation is a system designed and administered to be a fair and equitable process which balances the rights of employers, insurance carriers, and claimants.

4. **True** or **False**: A facility which generates a high number of low cost injuries is a better risk than a similarly sized facility which produces fewer but more costly injuries.

5. **True** or **False**: A corporate policy requiring workers' compensation claims to be reported to the insurance company within ten working days is a reasonable performance standard.

6. **True** or **False**: To minimize workers' compensation costs, employees should be fully trained in how to file a claim and be fully informed of their rights, benefits and entitlements.

7. The best place to position safety responsibility within the structure of an organization is in _____ _____.

8. **True** or **False**: Employee safety training is an effective way to improve the level of safe behavior in an organization.

9. **True** or **False**: An organization which places high emphasis on regulatory compliance can anticipate lower than average workers' compensation costs.

10. **True** or **False**: In safety, as in other performance areas, what gets measured gets done!

11. **True** or **False**: In a multi-location, multi-state operation, comparing workers' compensation loss costs location to location is an equitable way of measuring safety performance and motivating better results.

12. **True** or **False**: Workplace accidents are highly fortuitous events and, therefore, are difficult to predict.

13. **True** or **False**: A uniformly administered progressive discipline policy is an effective strategy for improving safe behavior in the work place.

14. In high-performance safety organizations, executives and managers recognize the number one cause of workplace accidents to be: _____.

15. **True** or **False**: An effective return-to-work program emphasizes timing — when an employee can resume job duties.

16. What percentage of workers' compensation injury lost time is due to physician-imposed restrictions?

    ❑10% ❑25% ❑50% ❑75% ❑90%

17. What is the organizational characteristic which most influences the level of workers' compensation claim costs incurred in an organization? _____

18. Employee fraud and system abuse account for what percentage of total workers' compensation losses?

    ❑5% ❑10% ❑20% ❑30% ❑>50%

19. On which day of the week are most compensable lost time sprain and strain type injuries reported in the workplace? (select one)

    ❏Monday ❏Tuesday ❏Wednesday ❏Thursday ❏Friday

20. By investing in a safety program, one can expect to reduce workers' compensation losses by _____ percent.

    ❏0% ❏10% ❏25% ❏50% ❏>75%

21. **True** or **False**: Returning injured employees to the workplace before they are fully capable of resuming their normal job duties ultimately increases loss costs in an organization.

22. **True** or **False**: To maximize workers' compensation loss reduction, it's important to focus priority on to the 'indemnity' (lost time) component of lost time claims.

23. An organization which achieves and maintains injury rates equal to its industry average can expect workers' compensation costs to_____ over the long term.

    ❏Increase ❏Remain Unchanged ❏Decrease

24. Ultimately, workers' compensation losses are allocated between the insurance carrier and the insured in the following proportions:

    Insurer = _____ percent

    Insured = _____ percent

25. **True** or **False**: It is common practice for insurance companies to set different levels of claim reserves on similar injuries submitted by different insureds.

WHAT'S YOUR ORGANIZATION'S LOSS MANAGEMENT IQ?

Record your answers in the spaces below.

Question	ANSWER
1	
2	
3	
4	
5	
6	
7	
8	
9	
10	
11	
12	
13	
14	
15	
16	
17	
18	
19	
20	
21	
22	
23	
24	
25	

See Occupational Hazard's December article: "STRAIGHT TALK" (Next Article)…About Safety, Workers' Compensation and Loss Costs for answers.

STRAIGHT TALK...ABOUT SAFETY, WORKERS' COMPENSAITON, AND LOSS COSTS!

LARRY L. HANSEN

Originally Published as December 1999 feature in' Occupational Hazards' magazine.

Unfortunately, in many businesses today, workers' compensation costs continue to stress organizations and pose very real threats to corporate profitability. In many respects, loss problems are the same ... only now the answers are different! Safety professionals tasked with leading their organization's response to these tests, are finding that traditional strategies (where to go and what to do) are no longer adequate to meet the challenges. We live in an era of harsh realities ... the realities of change:

1. **Business has changed** - We no longer compete on a local, regional, or national basis. Our business strategies must withstand the pressures of global competition. The Far East isn't that far; the Third World is catching up; quality is no longer an option; and 'human resources' have become the competitive edge worldwide.

2. **The business we do has changed** - We have changed from a manufacturing economy to a knowledge and service economy. Many of today's employees don't report to work daily; they 'work' continually in their minds.

3. **The business process has changed** - Business and academic leaders have taken us through a restructuring and re-engineering of American business. To be a world-class competitor, organizations now recognize the need to change from hierarchy to teams, from authority to influence, from control to empowerment, and from disciplinary approaches to reinforcing practices. 'B' schools now teach leadership of people in addition to the traditional curriculum of managing process and profit. We have literally turned our pyramidal organizations and mind sets upside down ... yet,

4. **Safety has NOT changed!** - The predominant models, methods, and mind sets which have guided the safety profession for decades remain unchanged. There continues to be a strong attachment to traditional safety strategies; the 3 E's of safety: enforcement, education and engineering.

Over the past decade, business leaders have clarified the 'critical factors' of business success. Tom Peters has helped us see the linkage between leadership and excellence; Stephen Covey, the importance of values and relationships; Peter Senge, the requisite for continual learning; and W. Edwards Deming, the need to focus on process and continuous improvement. But, above all, it has been Peter Drucker's insight that success is driven by attention to the critical 'middle lines' (loss and expense) of business. In his words, "Minimizing loss is the number one responsibility of management." It matters not how much revenue (top line) is generated if these dollars are diverted by loss and expense from reaching margin (the bottom line). Herein lies the challenge to today's organization. Enterprise 'trival' requires effective loss management!

Effective loss management, however, is not the sole province of the safety manager ... it is the task of every manager in an organization. The safety practitioner's challenge is to grow organizational loss management competency at all levels of their organization. What's needed is — 'straight talk' on those loss management factors critical to success: workers' compensation, safety, and management of loss and costs. Here are some of the answers from today's more successful loss management leaders:

WHAT'S YOUR LOSS MANAGEMENT IQ?

Assign four points for each correct answer you selected in the following exercise. After grading all answers, total your scores to derive your bottom line on loss management (total between 0% and 100%). Your total score and the four subject-specific subtotals will provide insight to your current organizational strengths and reveal loss management improvement opportunities.

	Your Answer	Points	
1.			
2.			
3.			
4.			
5.			
6.			Percentage
Subtotal 1 (Questions 1-6): Straight Talk About Workers' Comp			$\div\ 24 \times 100 = $ ____ %

7.			
8.			
9.			
10.			
11.			
12.			
13.			
14.			Percentage
Subtotal 2 (Questions 7-14): Straight Talk About Safety			$\div\ 32 \times 100 = $ ____ %

15.			
16.			
17.			
18.			
19.			Percentage
Subtotal 3 (Questions 15-19): Straight Talk About Loss Drivers			$\div\ 20 \times 100 = $ ____ %

20.			
21.			
22.			
23.			
24.			
25.			Percentage
Subtotal 4 (Questions 20-25): Straight Talk About Your Money			$\div\ 24 \times 100 = $ ____ %

Total (Subtotals 1-4): Your Bottom Line on Loss Management		%

SUGGESTED USES:

1. Administer the test to executives and senior managers in your organization to derive your leadership Loss Management IQ.

2. Administer the test to managers at different levels of your organization (supervisor, manager, executive) to derive your organizational Loss Management IQ.

3. Administer the test to various functions to derive your operational Loss Management IQ.

ANSWERS

1. **Buying Insurance** - Of the three recognized risk management strategies (risk transfer, risk finance, and risk avoidance), the most expensive and generally least effective strategy for dealing with workers' compensation risk is to insure it. Financing risk via insurance or other financial mechanisms is primarily a cost-plus proposition which adds friction costs to losses in the form of handling fees, taxes, and profit loadings. In essence, insurance finances loss problems out into the future rather than resolving them in the present. It's unfortunately all too common for operations managers to believe that workers' compensation losses belong to the insurance company; i.e., "it's their problem" (and dollars). This belief minimizes their concern with and involvement in problem seeing and problem solving. Bottom line -- losses unmanaged grow to problematic levels.

2. **False** - A 20% dividend on a workers' compensation insurance program doesn't attest to one's investment prowess. An insurance dividend is basically a retroactive return of pre-paid excess premium. It's money refunded to an insured (based on actual incurred loss levels). Insurance dividends are 'not taxable' ... as financial dividends. Sorry, but you don't accumulate financial wealth by insuring risk; you only save money by preventing accidents, and minimizing losses.

3. **False** – Workers' compensation systems are legislated to be equitable but, in practice, are administered to protect the rights of injured workers. Anybody attending a workers' compensation hearing will easily see the system is fairly balanced ... '50/50' — in favor of the claimant! Given the size, resources and financial strength of all other players involved, this should not be unexpected. Managers often protest what they perceive to be 'inequity' in the system. In reality, the odds actually favor the employer. The system, as administered, is the reasonable price employers must pay for the protection they receive prohibiting employee negligence actions under tort law. If these statutory 'no-fault' protections didn't

exist, negligence actions would prevail as the remedy for workplace injuries, and costs would be significantly higher than they are today.

4. **False** - A facility with a high frequency of small claims is the greater risk. The size (ultimate cost) of an injury is highly fortuitous. If a company generates a high frequency of accidents, eventually its 'luck' will run out and severe losses will be incurred. It's all a matter of the odds ... not if, only 'when.' There truly is a difference between 'bad luck' (severity) and 'bad control' (frequency) — Insurance regulators discount the first and penalize the latter. Frequency is manageable; severity is insurable.

5. **False** - Industry 'lag time' studies (date of occurrence to date of report) have confirmed that delays in reporting claims adds additional cost to injuries. A ten (10) working day reporting standard (up to 16 days including weekends) can add significantly to the ultimate cost of workers' compensaiton in an organization. In 1995, a Johnson & Higgins Quality Council, comprised of seven best practices companies, studied this issue in their own organizations and found that: "...for claims reported in under 20 days, each day a claim was reported one day sooner resulted in a 3% reduction in average claim cost." Studies by Kemper National Insurance also confirm the cost impact of late reporting. Their study revealed: "claims reported in excess of 30 days (from time of incident) were 48% more costly when compared to claims filed within 10 days." In claim reporting, time really is money ... compounded daily!

6. **True** - The cost of a workers' compensation claim is heavily influenced by a claimant's anxiety and fear of the unknown. It is critical to inform employees of the workers' compensation process, reporting procedures, and their rights and entitlements before injuries are incurred. Industry studies confirm that lack of information creates doubts and adversarial feelings that can lead to attorney involvement and ultimately higher claim costs. The Disney World organization significantly reduced litigation and the cost of their workers' compensation claims by implementing a progressive three-phase communications program comprised of: 1) informing claimants of their rights; 2) non-adversarial claim investigation; and 3) 'TLC" contacts by managers. When it comes to workers' compensation cost containment, honesty isn't only the best policy — it's also the least expensive policy.

7. **All Organizational Functions** - To maximize safety performance in an organization, safety responsibilities must be fully integrated into line operations and clearly defined in all management positions. A common error committed by many organizations in their efforts to control accidents is to assign safety responsibility to a staff function or separate 'safety department.' When this occurs, 'line managers' wipe their brows in relief — "don't have to deal with that stuff anymore!" The result: Safety becomes the safety department's job and those with the greatest opportunity to control those conditions and work practices which influence accidents (line managers), off-load their responsibilities. Safety must be 'everyone's job' ... and these jobs must be specifically defined. Typically when an organization is suffering loss problems, someone's not doing their job ... and that someone generally isn't the safety director!

8. **False** - Employee safety training, although one of the most common strategies, has not proven to be effective in improving the level of safe workplace behavior. Many safety performance studies, including a comprehensive study of back injury prevention training in the New England Postal Service, confirm that training has minimal impact on accident rates, average days lost, or average injury costs. The study concluded that what good training in fact did, was produce "better educated injured workers.' What truly impacts safe behavior, reduction in accident rates, and lower loss costs in an organization is consequence delivery; i.e., performance management and reinforcement practices. Employees do what they are reinforced and rewarded for doing ... not necessarily what they are trained to do!

9. **False** - There is little confirmed correlation between the level of 'regulatory compliance' and workers' compensaiton loss reduction in an organization. The reason for this is that compliance activities primarily focus on safeguarding facilities, conditions, and equipment while the true causes of accidents involve management values, organizational design, and work behaviors. Low workers' compensation costs are more an outcome of improved management processes than imposed regulation of physical conditions. OSHA's 1997 'strategic goal' to move toward impacting the 'culture' of organizations evidences recognition of the need to focus on results (the objective of management) rather than hazards (the objective of compliance).

10. **False** - Safety is no different than any other performance issue; measurement alone is never adequate enough to positively impact results. Many overweight people get on a scale every morning yet fail to do anything about dieting — (how well I know!). In business it's not what gets measured, but rather what gets measured, managed, reinforced and rewarded that gets done! Safety is an issue that gets heavily measured ... yet frequently ignored. The Quality Solutions Group of Nashville, Tennessee, in its research of industry practices, has concluded: "most organizations are data rich concerning their loss situations yet remain uninformed, and without understanding on how to use this data to solve their problems.' Improved results can only be generated by: benchmarking current performance, identifying improvement opportunities, setting goals, developing improvement strategies, assigning specific responsibilities, implementing an action plan, measuring progress and rewarding results. Hmmm..., if this sounds similar, it should, because it's the basic process of management! Safety can be managed ... but generally isn't.

11. **False** - Comparing workers' compensation losses location-to-location in a multi-state operation is an inequitable measure of safety performance. Workers' compensation losses are based on injuries, and injuries are highly fortuitous outcomes. Benefit levels and compensability rules also vary dramatically by jurisdiction. Using 'loss costs' as a performance measure creates an "apples with oranges' comparison. A more equitable measure of safety performance is the level of 'safe behavior' in an organization confirmed by on-floor observations. Another objective measure is the assignment of 'standard costs' to various injury types and applying these uniformly to injuries incurred at all locations. Under this approach — a fall, is a fall, is a fall — each is worth (on average) a standard amount regardless of actual cost outcomes. Using 'standard costs' assigns comparable values to accidents (the real target of prevention), not their fortuitous outcomes called injury!

12. **False** - Workplace accident rates, injury types and loss trends are highly predictable. Insurance industry historical data and Bureau of Labor Statistics rate trends clearly indicate accident trends and predominant injury types in our workplaces year to year. Our nation's 'all industry' lost time incident rate hasn't varied significantly over the past 20 years. The only truly fortuitous and unpredictable aspect of an accident is "when, where, and to whom it will occur' — How lucky do you feel?

13. **False** - Any form of punishment, including progressive disciplinary programs common in most organizations, will not increase the level of safe behavior in a workplace. Punishment can stop certain unsafe acts, but it does not increase the level of desired safe behavior. Only the use of reinforcement principles and practices will increase the level of safe behavior. Unfortunately, reinforcement strategies are poorly understood and applied by managers in most organizations. What improves work performance, including safe performance, is positive recognition and rewards. These, unfortunately, are the least utilized performance improvement strategies. Behavior consultants suggest that to improve human behavior, a 4 to 1 (positive to negative) ratio of consequences must be applied. Human resource studies have confirmed, however, that the ratio of negative inputs to personnel files compared to positive remarks is more like 8 to 1! — We've got some 're-thinking' to do.

14. In high-performance organizations, managers clearly understand that **accidents are caused by 'the system'** - designed, administered and maintained by management ..., not careless employees. In such organizations, there's clear acceptance that managers work 'on the system' while employees work 'in the system,' and that the system 'causes accidents' while employees 'sustain injuries.' According to Dr. Deming's teachings in statistical process control, less than 10% of undesirable process outcomes (including accidents) are due to individual behaviors (special causes). The majority (90% or more) are due to problems inherent in the design and administration of the process (common causes) - management controlled causes. Bottom line — an organization can never succeed in process improvement if it believes people are the problem.

15. **False** - An effective return-to-work program focuses less on 'timing' (i.e., 'when' an employee can return to work) and more on 'capability' ('what' an employee can do). The 'what' determines the 'when.' An effective return-to-work program emphasizes abilities — 'can-do's,' not disabilities — 'can't-do's!' "Organizations can expect an $8 to $10 savings on every dollar invested in a return-to-work strategy" reports Joyce Frieden in Business and Health Magazine. If employers really want to reduce workers' compensation costs, they must develop better 'can-do' attitudes.

16. **10%** - Industry studies confirm that only 10% of compensable lost-time is due solely to physician-imposed disability restrictions. All other reasons for 'lost time' are due to employer and employee controlled impediments to return to work, such as inflexible supervisory decisions, poor injury management practices, breakdowns in communications, and failure to make reasonable work accommodations. Researchers at National Rehabilitation Planners, Inc., identify these situations as 'co-malingering' — the

mutual actions of employer and employee which extend disability duration and impede early return to productive employment. Employees incur 100% of lost time; employers control 90% of it!

17. **That** organizational characteristic which most impacts workers' compensation loss is **"human relations. (Not the function–the practice!)"** Studies by the National Institute of Occupational Safety and Health and the Reliability Group, an organizational performance consulting firm, confirm a strong relationship between the level of safety (accident rate) in an organization and the strength of its employee/employer relations. In adversarial work environments, employees have a low tolerance for 'working through pain' and see the workers' compensation system as a 'way out' of an undesirable situation. Once out, they have little desire to return. This negatively impacts disability duration and loss costs. Ken Majejka sums it up best: "It's very easy for an employee in an unhealthy organization to come to the conclusion that work is highly overrated.' In workers' compensation cost control, smiles are worth big bucks!

18. **5% or Less** - While the total cost of fraud in the workers' compensation system has been estimated to exceed 20%, system abuse by employees has been confirmed to account for less than 5% of this total. The greatest part of the fraud problem consists of actions on the part of the other stakeholders in the system; i.e., physicians, lawyers, service providers, insurers and employers themselves, involving activities such as overstating financials, underreporting payrolls, falsifying reports, redundant billings, etc. Deep Throat's advice in All the President's Men to 'follow the money' appears applicable to the fraud issue in workers' compensation. As Tom Lynch of Lynch, Ryan & Associates has observed: "In workers' compensaiton, the money passes through many hands.' Employee actions are "part of the problem…,' but they are not 'the problem' in workers' compensation.

19. **"Monday, Monday, can't trust that day"** — Unfortunately, the Mama's and the Papa's had it right! Studies supported by the NCCI in the voluntary workers' compensation market and the US Navy relative to claim activity in their civilian worker population, confirm similar findings. These studies found a statistically significant difference to exist in the incidence of lost-time strain and sprain injuries reported on Mondays as compared to all other days of the week. Navy researchers estimated that up to 22% of those incidents may be non-work related occurrences imported into the workers' compensation system due to lack of deductibles and/or lack of indemnity payments in group health insurance. These findings suggest it's well worth keeping a keen eye peeled on Monday mornings. So, how was your weekend?

20. **Traditional Safety Programs Have Minimal Impact on Workers' Compensaiton Costs** - Studies by the National Institute of Occupational Safety & Health, and more recently by the Department of Energy have found that investments in 'traditional' safety programs (enforcement, education, and engineering) had minimal impact on accident rates and workers' compensation costs. Safety programs based on compliance education and engineering do not penetrate the deeper, systemic causes of accidents in an organization; i.e., management values, structures, and processes. With these key 'loss drivers' untouched, workers 'compensation losses continue. Department of Energy studies at select sites actually found an inverse relationship to exist. Those facilities which had the greater investments in

safety programs had higher workers' compensation costs — who woulda' thought! Further proof, safety is a management process, not a technical issue.

21. **False** - Returning injured employees to work via transitional work programs, compliments healing, maintains employee involvement, and strengthens attitudes and work relations. An effective transition program is a cost effective method of managing occupational injuries and minimizing costs, as it prevents the development of adversarial relationships which can lead to inflated injury costs. Rebecca Shafer Bruce, a workers' compensation cost containment expert, estimates that a well designed transition program consisting of four key elements: 1) work hardening; 2) job modification; 3) alternative duty; and 4) volunteer assignments (within medical restrictions) can successfully return 90 percent of injured employees within four (4) working days. Estimates by the NCCI and National Rehabilitation Planners, Inc., also suggest that effective injury management and return-to-work practices can reduce loss costs by 25 to 40 percent. To minimize loss, employers must develop a comprehensive answer to the simple question which lies at the heart of all lost time cases: "When do you want to go back to work?"

22. **False** - Data maintained by the National Council of Compensation Insurers shows that the medical and indemnity components of lost time workers' compensation claims each comprise approximately 50% of total claim cost. Over the past 10 years, the medical component has actually grown more rapidly than the indemnity. Given this 'near' 50/50 proportion and the escalation trend in medical costs, 'both' elements must receive equal priority in loss control efforts.

23. **Increase** - An organization that maintains 'flat' incident rates can expect significant increases in workers' compensation costs due to medical inflation, expansion of benefits, and growth in benefit levels. The total cost of workers' compensation has grown consistently over the past 20 years while national lost time incident rate trends have remained flat. The long term trends are clear — mediocre performance produces high costs.

24. **Insured** = 100%+ - For the vast majority of business, workers' compensation losses (over the long term) are fully absorbed by the insured. An insurance company collects premiums, provides services, administers claims, and distributes its insured's premium to service providers, medical professionals, claimants ... and Uncle Sam! Of course, reasonable handling fees, expenses, and profit loadings are added for providing these services — it's cost-plus, pay now or pay a lot more later! It's the insured's money! Here's an objective test of this financial reality: When's the last time your losses exceeded your insurance premium?

25. **True** - It is common for an insurer to set different loss reserves for similar claims occurring in different insured companies ... and no, they are not playing favorites! The reason for this is the claim reserving process. Where an insured does not have a proactive injury management process or return-to-work program in place to minimize loss cost, the insurance company must realistically expect to pay more

for injuries and, hence, they reserve accordingly. Reserving may not be an exact science — but it's not rocket science!

REFERENCES

"A Return-to-Work Program That You Manage," interview with Rebecca Shafer Bruce, *Risk Management Advisor* (A Bureau of Business Practice Newsletter) (Issue 309, September 1996), 1-4.

"Employee Oversight Teams Available," *Risk & Insurance* (November 1998), 20.

Borba, Philip S. and Sharon Eisenberg-Haber, "The Above-Average Incidence of Sprains and Strains on Mondays," *NCCI Digest* (1988), 3:51-62.

The Bureau of National Affairs, *Workers' Compensation Report* (December 5, 1994).

Carroll, Gwenneth, "In Search of Fraud Data," *The Lynch Ryan Report* (Vol. 3, Spring 1993), 3.

DeCarlo, Donald T., "Workers' Comp Update," *Risk & Insurance* (July 1994), 10.

Insurance Research Council 1993 Public Attitude Monitor Survey, *Risk & Insurance* (June 1994).

Katz, David M., "How Disney World Keeps Lawyers Out of WC," *National Underwriter* (November 13, 1995), 19, 28.

Kuhar, Mark S., "Workers' Compensation: Is Fraud Flying High?" *Occupational Hazards* (August 1994), 38-41.

Managed Comp, Inc., "Lost Time Unnecessary," *Risk Management* (July 1998), 9.

Medo, Dennis, "Rates Close to Endgame," *Risk & Insurance* (November 1998), 23-24.

Mitchell, Kenneth and Steven W. Leclair, Negotiated Disability: The Silent Partner in Cost Containment, National Rehabilitation Planners, Inc.

Negotiated Disability in The Health Care Industry: The Invisible Bond Between Worker and Employee, National Rehabilitation Planners, Inc. (1993).

NCCI West Coast Forum presentation (November 16, 1998).

Occupational Safety and Health Administration, OSHA Facts, www.osha-slc.gov/OshDoc/OSHFacts (1998).

Samuel, James, "Workers Comp Fraud: Spotting Illegitimate Claims," *Ohio Bureau of Workers' Compensation* (Fall 1996), 11, 13-14.

Shepher, Steven L. and Bonnie J. LaFleur, "The Increased Incidence on Mondays of Work-Related Sprains and Strains," JOEM (July 1996), 38:7.

Tillinghast-Towers Perrin, Reality Testing: Assessing the Performance of Workers' Compensation Cost-Management Initiatives (Survey Report), 1995.

Wausau Insurance Companies, Early Return to Work: Something Special (1996).

Wausau Insurance Companies, Wausau's Managed Disability Program: People at Work (A Proactive, Comprehensive Approach to Reducing Time Loss and Work Absences).

Weinstein, Mindy, "Crafty Claims," *Risk & Insurance* (June 1994), 29-30.

Wojcik, Joanne, "Trimming Comp Costs by Zeroing in on Fraud," *Business Insurance* (May 2, 1994), 22-23.

THE REAL DRIVERS OF SAFETY

Safety At Work interviewed Larry L. Hansen of the United States Company, L2H. L2H is a Performance Improvement training and facilitation company with capabilities in the advanced strategies of safety excellence. L2H provides professional development conferences, leadership training, and performance improvement seminars, which address excellence strategy and organizational change.

These strategies and the excellence model they comprise are described in "The Architecture of Safety Excellence" published as the May 2000 cover/feature of Professional Safety magazine. Further information is available at www.L2HSOS.com

Larry will be presenting a paper on "The Architecture of Safety Excellence" at the Safety In Action conference that begins in Melbourne Australia on March 30, 2004.

SAW: Larry, could you provide us with a bit of an introduction to your company L2H?

LLH: As a preface to this interview, I'd like to offer some upfront comments. In the safety profession there is a vocabulary issue that needs to be addressed. I will talk about three important words – Safe, Safely and Safety. These need explaining, as they are really critical. Safe is a value; it's a basic belief in how we do business. Safely, is a process, it's the methods we use to do business. Safety is a measurable outcome of the business process. While we go through this interview it will be useful to keep these concepts in mind, Safe - a value, Safely - a process and Safety – an outcome.

I believe that the biggest impediment to 'Safe' (the outcome) is our profession's current belief that safety (the process) is something that needs to be managed very differently and separately from the way we manage business in the rest of the organisation. We tend to talk about the need for a safety program, safety management, safety leadership, and a safety culture. I don't believe that that is a very successful agenda. Safety is the outcome of having 'Safe' as a value in a well-managed business. We need to focus on the core elements of the business process, the drivers of the outcome called Safety.

For a bit of background to L2H, my education is in industrial management, my background is in manufacturing, management and casualty insurance, risk and loss management. I spent 34 years in that profession with a couple of companies. There have been two careers in that period, the first career, (18 years), was as a very traditional safety practitioner. I would talk to companies about safety regulation, and the need for safety programs. Over time, as I did this, I saw a couple of patterns develop.

I was never really given access to senior leaders in my client companies; they were always out to lunch or out of the building. When I was delegated to a mid-line manager, after ten minutes of conversation he would get a glazed expression on his face, and you could read him thinking "when is this guy going to be done?"

The other thing I noticed was that after multiple visits to these clients, the problems never got solved, they were only relocated. We would fix, perhaps, a machine guarding problem in the East wing of the plant, and on our return visit, the same problem would exist in the opposite wing.

I began to question whether the strategies I was using at the time, the traditional strategies of the safety profession – engineering, education and enforcement of regulations – were appropriate for achieving high performance in business.

SAW: Safety seems to be evolving into a multidisciplinary process of risk management. Is this a good thing?

LLH: Risk Management has a real opportunity to be a catalyst – a function that could bring together both the line and staff elements of an organisation to effectively assess, evaluate and mitigate risk.

SAW: The risk management focus of major companies is usually on finance matters. How can we piggyback on the corporate accountability issues to get safety in the attention of senior managers?

LLH: The recent Enron collapse in the United States lead to a number of laws being passed that call for higher standards for corporate accountability – record-keeping, bookkeeping, financials and such. That has cast a spotlight of opportunity on business as far as safety and risk management goes. One of the outcomes of these additional laws is that when companies occur losses such as workers' compensation losses, these are to be valued to their ultimate cost. The books need to reflect the true exposure of these situations. What we are starting to see is that these are really large numbers, numbers that are attracting corporate executive attention perhaps more than in the past. Risk and safety managers can step forward now with good strategies to minimise these large costs in organisations.

SAW: Has this resulted in financial maneuvring's or has this spurred preventative action?

LLH: Both, quite frankly. My experience suggests that most companies still mismanage risk in the workplace, they are not doing the right things so they are not getting the right results; they are choosing to finance the cost via insurance mechanisms or other risk transfer mechanisms into the future. The more enlightened and progressive thing companies have seen the opportunity to address risk, not so much in the finance matter, but recognising that good risk management and safety management is the way to improve operating processes and to make the company stronger.

SAW: Safety seminars on new hazards always generate the question "can we insure against this hazard?" It seems to be a major hole in business attitudes towards Safe.

LLH: I hear similar things. We talk about exposure to loss and it seems that the deeply-seated mindset is not to think in terms of how do we prevent that but more of how do we finance that problem into the future. The bottom line is whenever I get into dialogues on that issue in seminars or with clients, there is a real need to help people realise that financing a problem makes the problem far more expensive. We delay a large bill that we have to pay.

Corporate executives need to realise that it's a lot easier and cheaper to secure the barn door than chase the runaway horses. When you get a chance to discuss the cost of preventing problems or pre-empting problems, versus financing them into the future, people see that the least expensive and the better way is the preventive way.

SAW: Do the government safety regulators in the States operate in this way or are they compounding this misperception of companies?

LLH: Regulation hasn't been very effective in this country. You can't mandate Safe into a work environment. It is not an effective strategy for improving operations.

SAW: In many countries safety law is meant to be based on Common Law and common values, it evolves to increase penalties and struggle with enforcement to bring back the values it was meant to be reflecting in the first place. Is there a similar cycle in the United States?

LLH: Regulations are passed in the States and then larger fines are applied but I don't think regulations and legislation drive the decision making of corporate executives. Good business sense does. Seeing the dollars and cents of a good proposition does. When safety and risk management is put into the good business vernacular with hard facts and dollars and cents, corporate executives, most are bright individuals, get it quickly, go forward and are very successful.

I still see the large mass of companies being regressive in their thinking, but larger and more progressive companies have seen where it makes business sense, do the right things and they are getting far superior results.

SAW: How do you suggest your clients keep up with some of the, supposedly, new occupational hazards like

bullying and stress?

LLH: I don't think that the hazards are new. I think what happened is that the environmental changes that allowed those hazards to manifest into financial compensable situations got people's attention. I have found, generally, not much problem with corporate America being aware of their exposures. The media does a fantastic job of publicising events. Most organisations have somebody in a staff position that overlooks personnel or risk matters and keeps decision makers informed on the need for action. It's an evolutionary process.

SAW: One of the comments in one of your recent articles caught my attention. You proposed setting a goal of 300% increase in the reportable incident rate for a calendar year. If you mentioned that at a safety conference there would be a sudden intake of breath and maybe laughter. How successful can that type of attitude and goal setting be?

LLH: I have a business partner, Dan Zahlis, who is a very progressive thinker in understanding that what business wants is to mitigate and reduce its cost of loss. From a business perspective, senior executives want to reduce cost; they see things in terms of dollars. The point you mentioned came from Dan's experience. He worked for a large company, and he his counterparts from other divisions were called to a corporate conference where each had to present their plans for improving their loss experience in the upcoming year. These types of meetings are often very competitive with everybody putting their best foot forward. When it was Dan's turn he said he was going to increase the number of Reportables. He was told he couldn't do that so he said, "watch me.' But he also said that in doing so he would reduce the loss cost associated with the reportables. That happened. He significantly reduced the loss cost.

His premise, and the whole truth behind it, was that in any organisation at any given time reality exists, things are happening, people are being injured, and events are occurring. Because of all kinds of pressures and stresses, the real data is being oppressed and being understated in the workplace. If you don't know about it, you can't solve it. Dan's intent was to take the lid off Pandora's box, let the truth come forward, motivate people to tell what the actual problems were so that they could see the causes, and solve the problems once and for all. By increasing the numbers reported, he established a trust level in the workforce that revealed and solved problems and reduced loss cost.

SAW: It sounds like companies are embarrassed about safety. What has made us embarrassed about our safety performance?

LLH: The deep-seated inherent belief of managers is that the safety problem, the injury problem, is an employee issue. In other words, employees are the problem, not us. The whole Quality field should have taught us that that attitude is all wrong. The majority of things that are defective in an organisation, defective outcomes including the production of accidents and injuries in the workplace, are not the fault of stupid people; they are the fault of the system that Management develops. In the safety world, we have not adapted

137

the concepts of Total Quality Management into safety management.

SAW: How does that fit with some of the recent moves in the Behavioural-Based Safety (BBS) field? Where does that fit in changing the values of Safe?

LLH: When talking about behavioural-based safety, caution is required. I have worked under the umbrella of what is considered behavioural-based safety. I know some quality people in that discipline. If you speak to the more progressive thinkers on the BBS movement, these folks will say that we observe workers in the workplace to find at-risk behaviours. But we don't want to just find those behaviours and say to people, "don't do that" or whack them for doing that. After we find at-risk behaviours we ask, "Why are you doing it that way? Why are you acting in an at-risk way?" By asking that type of question we generally find systemic reasons that are forcing people to do things the way they do. That's the ultimate goal of behavioural safety – to observe people, identify at-risk behaviours, ask why they are doing it that way, be lead to the systemic causes and then to solve those causes. That is good behavioural safety.

Unfortunately many or maybe most of the behavioural safety implementations or attempts of companies stop at finding the at-risk behaviour and punishing the employee. That is not an effective strategy. Behavioural safety as conceived in its purist form in a valid way, it is not being implemented appropriately. As a result it does not last; it tends to generate employee resistance and long-term problems.

SAW: One of the issues you will be meeting at the Safety In Action conference in Australia at the end of March, is that the Victorian OHS legislation is undergoing a review, and the report may be out in time for the conference. One request that comes up frequently in the submissions to the review is that we need to increase the responsibility and accountability of workers over their own actions. I hear that type of request often in the context of behavioural safety. Do you think that workers value safety?

LLH: I sure do. In the many years that I have managed risk in the workplace, I had occasion after occasion after occasion to sit in frontline employee focus groups. I find that these people are intelligent and know what the issues are. There is nobody that I am aware of who wants to amputate his or her hand with a bandsaw. I am a firm believer in Dr. Deming's belief that a majority of accident outcomes are systemic in nature. The employees are the triggers of accidents, the employees do things at a certain place and a certain point in time that can lead to an injury conclusion but they are not the cause. The systemic and organisational reasons that employees do those things are the causes.

SAW: How do your programs translate away from Corporate America to smaller organisations and other countries?

LH: The concept of operational excellence and that values drive performance is a universal concept. It is easier to get people to understand, implement and apply in smaller organisations because they are less complex. The larger an organisation becomes, the more impediments are built into their own functioning.

Anything, including safety, is harder in a large organisation.

In terms of global relevance, I communicate with South America, Far East, Europe, Canada, US. The cultural drivers of employee performance are the same everywhere. The value systems are pretty uniform.

SAW: What do you hope to gain for yourself from attending the Safety In Action conference?

LLH: I plan to make comment at every session of the conference I will be involved in that I am there to learn. To learn what works and what is successful in that part of the world. I know that a lot of what we do in the States is not working but I think there is an opportunity for storytelling and experience sharing. If we can link these two parts of the globe more effectively, we will find what is right to do. I am not familiar with Australian approaches to safety, the regulations, and value sets but I do know that a lot of what is being done in the States is not an effective strategy or approach. I will bring some stories on what is not working here and stories on what I have seen work. I'll share these and hopefully get confirmation of common denominators.

SAW: You talk about 'safety excellence.' Does this equate to 'best practice?' What's your response to senior managers and safety professionals who talk about the best practice jargon?

LLH: I am glad you mentioned jargon because, many times that's exactly what best practice is. In most of my discussions where best practice comes up, it has been used as jargon – words upon which there are not a whole lot of understanding. In the States when the concept of best practice rolled into corporate America, everybody took whatever risk management tasks they were doing and labelled them as best practice. Does that make it the best thing to do?...the thing that everyone should do? I don't think so. Best practice has an orientation to be jargon rather than effectual.

Best practice deals with efficiency – how well are we doing things? Safety Excellence deals with effectiveness – doing the right things. In the States, over time, and through discussions I have had with people in Australia and New Zealand, safety in a very traditional way is being viewed as a matter of regulation, as a matter of the employee being the problem, and as a matter of engineering solutions. Those strategies have been used for 60 years with not too much success. Safety Excellence says we need not abandon those strategies, but realise that they are not the strategies of excellence. We need shift our focus to ways of making those things more successful such as focusing on organisational culture – the values that make safety important or not important in that company, the leadership skills – what leaders do to make safety something the way we do business or something that is ignored in business, and where we place the function of safety in our companies. Do we create roles and relationships where line and staff create partnerships to develop a safe process, or do we create organisations that place people in conflict – production vs. safety, maintenance vs. production, line vs. staff? Safety Excellence strategies focus on culture, leadership, and organisation, not so much best practices that ask what is the best way to do regulation, what is the best way to train people, what is the best way to do engineering.

139

THE TOP TEN WAYS TO IMPROVE SAFETY MANAGEMENT

SANDY SMITH, EDITOR OCCUPATIONAL HAZARDS

Originally published in the December 2004 issue of Occupational Hazards

A panel of safety experts offer their advice to improve the management of the safety process and foster leadership.

Safety management can be a touchy topic. Disagreements abound: Should companies go the route of behavior-based safety, or follow a systems approach? Should safety be management-driven or employee-driven? What metrics should be used to assess the safety process?

We recently spoke to a number of experts in the occupational safety and health field. Though some of their recommendations seem at odds with each other and they approach safety from different perspectives, two themes reverberated throughout the comments. The first was the need for safety leadership, not just safety management. The second was the need to incorporate safety into the organizational structure of the business and not treat it as a separate function.

Here is what they had to say:

1) Recognize the difference between managing and leading – Tom Krause, Ph.D., CEO of Behavioral Science Technology Inc.

Krause, who along with John Hidley, M.D., pioneered the application of behavioral science methods to safety with Behavioral Advanced Performance Process (BAPP) technology, says organizations can improve safety management by recognizing the difference between managing and leading, and should place more emphasis on leading. "By managing, organizations make things happen. It's a linear, practical function," says Krause. "By leading, organizations show employees why safety matters, why they should be motivated to get behind it and want to do it."

He has found that most companies are very strong on the managing side. They know how to make things happen. On the leading side, however, "there are usually significant opportunities" for improvement, says Krause.

"It is so pivotal to understand the difference between managing and leading, especially at the senior level," says Krause. "If senior leadership gets it right, then the culture will change. If senior management doesn't get it right, then everything else is like swimming upstream. It's a struggle."

2) Integrate all aspects of the safety 'program' into a single comprehensive management system – Richard Fulwiler, Sc.D., CIH, president, Technology Leadership Associates

After retiring as global director of health and safety for Procter & Gamble with 28 years of service, Fulwiler became president of Technology Leadership Associates, which specializes in increasing individual effectiveness and improving organizational capacity in the safety area. He advises, "Don't have a number of stand-alone programs such as lockout/tagout, job safety analysis, behavior-based safety, confined space entry, etc. Instead, be sure that all elements of the safety program are integrated into a single management system that is owned by line management."

Fulwiler suggests developing and deploying a 'before-the-fact' metric for measuring safety performance that leads to predictable and desired results. "In other words, focus on a system that prevents problems versus just solving problems, much the way fire prevention (before the fact) is superior to fire fighting (after the fact)," he says.

3) "POLICE" your safety program – James "Skipper" Kendrick, CSP, president, American Society of Safety Engineers and manager, Industrial Safety & Hygiene, Bell Helicopter Textron Inc.

"Management does not need to do anything special to improve the integration of occupational safety, health and the environment into the business realm," says Kendrick. "Safety needs to be managed at the same degree as every other aspect of business. The same amount of management skill and effort needs to be applied to safety as with quality, cost, schedule, production, etc."

He suggests using the acronym POLICE – Plan, Organize, Lead, Inspect, Correct, Evaluate – to focus on safety management.

He offers this description of POLICE:

Plan – "Plan for safety, health and the environment (SH&E) in everything you do."

Organize – "Organize so that SH&E is an equal player with all other business entities."

Lead – "Lead by example, walk the talk. As one executive said to me recently, 'If you want to know my position on safety, watch my shoes.'"

141

Inspect/Investigate – "Look for hazards and press for a timely corrective action. Investigate why the conditions exist, look for root cause and again drive for mistake-proof solutions."

Correct/Coach/Commend – "Correct items/situations found in a timely manner. Actively coach for safe performance. Commend safe activity and performance."

Evaluate – "How is the system/program functioning? Evaluate, develop solutions and press for continued improvement."

4) Integrate safety into the processes of the business – Donald J. Eckenfelder, CSP. P.E., principal consultant, Profit Protection Consultants

"Don't have anything beginning with the word 'safety'," counsels Eckenfelder, author of Values-Driven Safety. "Don't have safety meetings, safety processes, safety committees." In other words, have production meetings, manufacturing processes and work committees, because otherwise, "you isolate safety and when it's treated separately, it tends to be subordinated," he says.

Eckenfelder, a past president of the American Society of Safety Engineers, says one characteristic shared by companies that have world-class safety is "they tend not to have safety professionals or safety meetings. Those are integrated into the business process. The outcome is that the responsibility for safety is shared by everyone."

Those companies also care as much about employees' off-the-job safety as on-the-job safety, and they focus on the process, rather than the numbers. "If you ask someone at a company with world-class safety about accident and injury rates, often, they can't answer you because they don't know," says Eckenfelder. "But they can speak to you at length about the integration of safety into the business."

5) Identify clients and internal customers who see value in your services and make those customers your boss – Larry Hansen, CSP, ARM, principal of L2H Speaking of Safety Inc., and Dan Zahlis, president of Active Agenda, Inc.

No one will accuse Hansen or Zahlis of being shrinking violets when it comes to their opinions about safety management. Noting that almost 80 percent of safety professionals polled in a recent readers' survey indicated they're dissatisfied with their jobs, they pointed out: "What is buried in these statistics... is that safety professionals complain they aren't supported, aren't listened to, and haven't received their due respect from senior management. Yet, in the next breath, they openly acknowledge that they haven't expanded their knowledge base, explored new strategies, dug-in their heels of conviction, and are fearful of 'pushing back' in their organizations. They simply accept their plight and rationalize their boss's view of them as window dressing, or a necessary expense (for now), and then have the nerve to claim they're underpaid!"

Hansen, the creator/author of "The Architecture of Safety Excellence" and author of ROC Your Organization: Fifty-Two Ways to Instigate Radical Organizational Change for Safety Excellence, www.L2Hsos.com and Zahlis – the creator of "Active Agenda," a Web-enabled automated risk management data technology, and author of The Hidden Agenda and CAUTION: Beware OSHA Statistics www.ActiveAgenda.net – recommend those 80 percent of safety professionals who are dissatisfied with their jobs "identify those clients and/or internal customers that see value in your services, and concentrate your time and energy on them. Do so, even if your boss gets pissed... and do so even more if it gets you fired. Make your customers your boss, rather than the Accounting Department, corporate rules trolls or the corpse that signs your paycheck (it's likely a rubber stamp anyway)."

6) Don't make safety a 'priority' – Michael S. Deak, corporate director, Safety and Health, Compliance Process Safety and Fire Prevention, E. I. DuPont De Nemours & Co.

World-class safety performance – and safety management – requires leadership from the CEO and every other employee, says Deak, who celebrates his 40th anniversary at DuPont in May 2004.

"At DuPont, safety is a core business and organizational value. Don't talk about safety as a priority," Deak counsels. "Think back to Sept. 11, 2001. The priorities of most organizations changed. At DuPont, our business priorities changed, but because safety, health and environment is a core value, it didn't change. It's going to be there next year, it's there now, it was there last year."

Since priorities can change, organizations that include safety as a priority create a culture of people who hide out in foxholes, Deak believes. "They hide and think, 'This too shall pass.' They don't participate, because eventually, there will be a different set of priorities."

7) Management commitment and leadership and employee participation are key to safety management – Neal M. Leonhard CIH, CSP, manager, Safety Systems, MeadWestvaco Corp.

"With the increasing emphasis on social responsibility, many Fortune 500 companies are focusing on improving their injury/illness prevention systems," says Leonhard, "Collectively, they are facing the same questions that have been asked since the industrial revolution. What makes a difference in safety, particularly in today's global environment? Management commitment and leadership can make a difference in achieving sustainable results in injury/illness prevention."

MeadWestvaco has communicated a policy to set clear expectations of the current management team regarding safe and healthful work practices and conditions in the business unit. Goals and objectives are established and aligned throughout the organization to drive improvements in safety performance. Leonhard says MeadWestvaco managers are as conversant with safety as they are with other business issues such as production and quality.

"Management leads consistently with a philosophy that all occupational injuries and illnesses are preventable. Management's commitment to safety excellence is demonstrated through visible leadership, such as regular participation in safety activities, and encouragement of employee participation in safety efforts," he adds.

At MeadWestvaco, management provides for and encourages "meaningful employee involvement in the accident prevention system," he notes. "Employees are given the opportunity and are encouraged to provide input into the design and operation of safety processes/programs and the decisions that affect their safety and health. Employee input is valued and used."

8) Take a rational, disciplined approach to safety – Donald Eckenfelder

"Safety should not be an emotional subject any more than anything else in business. Take a rational, disciplined approach," suggests Eckenfelder. Many companies try to play on the emotional aspect of injuries, i.e., the impact an injury could have on the lives of employees and their families. "That might work for a while," he says, "but not for long and not consistently."

He suggests devoting energy to finding the root cause of anything that goes wrong in a disciplined way. And don't be misled by red herrings: "What's investigated is usually the symptoms of the problem when the real problem is the culture," he notes. "The root cause of the Columbia shuttle crash wasn't stuff flaking off the outside of the shuttle. The root cause was the safety culture at NASA."

9) Make everyone accountable for safety – Michael S. Deak

"Line management has to be held personally responsible for safety. That means they can be held accountable," says Deak. "In successful businesses, leaders are graded not only on a financial scorecard, but on their ability to integrate safety into the business process. At DuPont, safety is a core value, and core values are integrated into the day-to-day operations of the business."

Employee engagement is key here, he adds. And by 'employee," he means all employees: "Everybody from the CEO down is an employee. Every one of our 79,000 employees at DuPont is engaged in the safety management process. If you don't engage all employees, you don't have a prayer at becoming a world-class company, in safety or in business," Deak insists.

He says DuPont engages line employees in safety through frequent communication of core values, by including them in incident investigations and by facilitating their participation on audit teams, among other things.

In addition, line employees see management 'walking the talk,' he adds, and that probably has the most positive influence on their safety performance.

"Very seldom in my 40 years at DuPont have I seen people not do what they think the organization wants them to do. If they see a manager carrying two armloads of boxes up a flight of stairs without holding onto the handrail, it doesn't matter if we tell them 100 times to hold onto the handrail. But if they see the manager holding onto the handrail, that makes all the difference."

10) Get results or get fired – Larry Hansen and Dan Zahlis

"Get results or get fired, because it's only results that count," say Hansen and Zahlis, "and it's only results that will ultimately free your soul, and enable you to discover the real opportunities to make a difference in this profession, in your life and in the lives of those entrusted to your care and responsibility!"

A UNIVERSAL MODEL FOR SAFETY X-CELLENCE.

SECTION 6

Core Values & Guiding Principles

"The content of your character is your choice.
Day by day, what you do is who you become.
Your integrity is your destiny -
it is the light that guides your way."

~ Heraclitus, Greek Philosopher

"You can't manage yourself out of situations
you behave yourself into."

~ Stephen Covey

The Universal Model of Safety Excellence

Safety Leadership

Values

Strategy

A QUESTION OF EXCELLENCE:

VALUES
Do we believe safe operations are important to the ultimate success of our mission?

COVENANTS OF THE ROSE

LARRY L. HANSEN

Originally Published June 2004 cover/feature of' Occupational Hazards' magazine.

'A BELIEF SYSTEM FOR 'REDEFINING OPERATIONAL SAFETY EXCELLENCE'

I recall a telephone conversation I had some time ago with a long time friend, a highly respected peer, and perhaps one of the most progressive thought leaders in our profession today. He spoke of the challenges (and frustrations) he faced as the 'Safety Executive' of a major fortune company, and how difficult it was to function at that level, as the issues were so unique and complex. At that level of practice, success is all about diagnosing 'cultural problems', and creating 'transformational change'…in values, structure, and leadership practices. He acknowledged that there were few who could offer insight and credible advice, and was kind enough to grant me such privilege, an honor for which I remain most grateful today.

At that time, I was pursuing a career in safety management consulting, and had discovered that he was indeed correct. There were but few (in my own, or my client's organizations) who were willing to take the organizational risks, confront the status quo, instigate organizational change (for the better), or incur the political battle scars of 'doing the right things'. When I met such a person, it was an amazingly refreshing experience. They all had names…'D. A.', Keith, Dan, Ray, Paul, Pierre, Kelly, Hank, Don, (a lengthy list now), but more importantly they had unique minds…free spirits that refused to be shackled by the conventional 'myth-conceptions and wiz-dumbs' of the trade, a/k/a wrong headed thinking that impeded progress in the right direction!

Over time, these individuals became a loosely knit 'network of excellence'; a network that has grown in number, as I have interacted with many in this trade across four continents. Those who 'got it', were far and few between, however they were rich in knowledge and experience, and as time would prove, were those who ultimately would be successful. It is from this enlightened group, some of who had such strong convictions that it cost them their jobs (an event they now look back upon with pride) …that the 'Royal Order of the ROSE' has evolved, and the 'Covenants of Excellence', a belief system for safety excellence has been forged.

The 'ROYAL ORDER OF THE ROSE' recognizes those in this profession who have stood tall to the challenge, stiffened under bureaucratic pressure, fought the good fight; proudly display their scars of conviction, and keep on ticking! The 12 organizational 'truths of excellence' that follow are an evolved set of core beliefs forged over 30+ years of observation, research, practical experience, and interaction with these progressive thinkers…on what works, and equally as important, what does not, in generating safety excellence results. These foundational truths comprise those critical beliefs which, when embraced by an organization, can enable it to 'Re-define Operational Safety Excellence', and achieve operational safety success. These 'Covenants of Excellence' are as follows:

Covenant #1 - 'Safety Excellence' is not about preventing accidents, a/k/a 'end-of-pipe' reactive activities devoted to finding process problems, physical hazards, and at-risk work practices via compliance inspections, job observations, and progressive discipline policies.

Safety Excellence is all about 'proactively' designing, aligning, and improving the operational process; i.e., identifying upstream opportunities to minimize exposures, defects, and variances by implementing organizational changes that improve operational effectiveness, increase productivity, and minimize loss cost.

Covenant #2 – At-risk employee behaviors do not cause accidents. Accidents are caused by the reasons for 'at-risk' employee behavior—these 'causes' are organizational and cultural in nature…issues of management and leadership.

At-risk behaviors (people) are not the problem; all behavior is caused. To achieve excellence, seek, expose, understand, and remedy the 'good reasons for poor performance', the problems embedded in the management systems and leadership practices of the organization. People don't behave as they believe…people behave as they believe their bosses, the level of fear, the performance metrics, the compensation structure, and the recognition and reward systems want them to behave…and then they go home and change.

Covenant #3 - Accidents are not the problem; the problem(s) are the problem.

Accidents are not unplanned, unforeseeable, fortuitous process outcomes. Accidents are patterned and predictable performance symptoms; the final visible evidences of systemic failings and organizational deficiencies. If there are accidents in an organization, there are systemic organizational problems. To achieve excellence, focus on the systems, not the symptoms of accident causes.

Covenant #4 - The 'business process', defines, drives, and ultimately determines all business outcomes, of which safety is but one.

All organizational outcomes (good or bad—complying or non-complying—on time or late—profitable or unprofitable and safe or unsafe) originate from common headwaters, and are delivered by a common system. Accidents are an organizational design, administration, and maintenance issue; seek to remedy organizational, managerial, and leadership errors of co-mission and omission.

Covenant #5 - Employees work 'in the system'; managers work 'on the system'; the system produces accidents; employees sustain injuries. Manage one to minimize the other.

Accidents, and their causes, correction, control, and costs, are the responsibility of mid-line managers and topside leaders, not the fault of front-line workers. It is unrealistic and unreasonable to ask people to fix a system that they do not design or control. The functions of 'Planning, Organizing, Directing, Controlling, Measuring and Monitoring' work and process are (by definition) components of the 'Management System'.

Covenant #6 - To increase the 'bottom line', managers must effectively manage the middle lines, of which, the cost of L.O. S. S. (Lack Of Safety Strategy) is significant. Safety excellence is ultimately measured below the line…but must be proactively managed above the line.

Minimizing, and at every opportunity eliminating, operational loss, cost, and expense is a manager's primary responsibility…this is also called 'productivity'!

(**Note:** This is, Peter Drucker's position! The only things a manager can manage, i.e., plan, organize, direct, control, and monitor are the middle lines…losses, costs, and expense… a/k/a accidents, injuries and claims.)

Covenant #7 - Safety performance is a clear, accurate, and reflective measure of the strength of an organization's leadership values, management competencies, and operational systems.

"How leaders lead, determines how managers manage, and how managers manage, directly impacts how employees perform…including Safe vs Unsafe." All organizational outcomes (including negative ones) are a function of: leadership (values), management (practices), and organization (structure and systems). These are strategic business issues; and form a scorecard of operational effectiveness.

Covenant # 8 – A 'core truth' is deeply embedded within the policy and value system of all organizations, which repeatedly generate poor safety performance, and it is: 'People don't count—every other number does!'

It's easier (more expensive, but easier) to neglect workplace health and safety than it is to manage it effectively. Line managers in under-performing organizations follow the Law of Least Resistance: "Given the opportunity to do nothing, most will!" Excellence companies don't abide by this law!

Covenant #9 - Achieving safety excellence is requisite upon measuring and managing the right things…most organizations fail to effectively do either.

Measurement based solely on loss outcomes (i.e., accidents, injuries, claims and loss cost) is regressive practice, and impedes an organization's performance. Excellence requires measurement of all four-performance parameters: (predictors, leading indicators, trailing metrics, and results), and at all four levels of organizational responsibility: (executive, manager, supervisor and front-line.)

Covenant #10 - Safety must never be the responsibility of a 'staff' function; safety must always be the obligation of 'line' managers.

Excellence is achieved when an organization builds unified business systems in which safety is integrated into core operational processes, line and staff roles and responsibilities are effectively designed and aligned, and collaborative partnerships across the organization replace sub-optimizing inter-functional competition. Eighty percent of work performance in an organization is shaped not by 'the written rules' (safety program developed by staff), but by 'the unwritten rules'--the consequence delivery system (actions or inaction) administered by line managers.

Covenant #11 – Discipline, the most common performance management technique employed by managers, does not increase the level of safe employee behavior in a workplace. Only reinforcement, a far 'used-less' management practice does.

'Whack a Mole' sometimes produces winners at carnivals and State Fairs…but is always sub-optimizing in the workplace. People strive to achieve higher levels of performance when they 'wanna' -- are motivated to do so. When they are told they 'gotta…or else', people divert the greater part of their energy away from being productive to one single objective -- getting even! And, they generally will, without your having the slightest clue. If some morning your coffee tastes 'really strange' don't say, I didn't warn you!

Covenant #12 – Excellence is attainable…NOW! Any organization can be a 'Safety Excellence' organization; most however, will defer to 'L.A.M.E.' (Lethargic, Antiquated, Mediocre, Externally-focused) excuses…and whine a lot!

Excellence requires two critical elements: 'knowledge and hard work'. In safety, most organizations resist developing the first, and are unwilling to expend effort on the second. In business, (and safety), there is no such thing as a 'quick fix'. However, when strategy is sound, leadership is strong, resources are adequate, and efforts are meaningful...there is great potential for 'rapid returns'. Unfortunately, most organizations have an antiquated 'No Returns' policy!

Of course, for those not willing to accept these covenants, make the necessary commitment, or implement the transformational change requisite to success, there are still a number of options available. Among the more common are:

- Increase the annual Workers' Compensation Premium budget.

- Hire more lawyers.

- Lower profit margin projections!

Mediocrity indeed has its price!

COVENANTS OF THE ROSE

' A Belief System For Safety Excellence'

COVENANT #1 - 'SAFETY' IS NOT ABOUT PREVENTING ACCIDENTS; 'SAFETY' IS ALL ABOUT IMPROVING THE PROCESS.

COVENANT #2 - EMPLOYEE BEHAVIORS DO NOT CAUSE ACCIDENTS; ACCIDENTS ARE CAUSED BY THE REASONS FOR 'AT-RISK' EMPLOYEE BEHAVIOR.

COVENANT #3 - ACCIDENTS ARE NOT THE PROBLEM; THE PROBLEM(S) ARE THE PROBLEM.

COVENANT #4 – THE 'BUSINESS PROCESS' DETERMINES ALL BUSINESS OUTCOMES, OF WHICH SAFETY IS BUT ONE.

COVENANT #5 - EMPLOYEES WORK 'IN THE SYSTEM'; MANAGERS WORK 'ON THE SYSTEM'; THE SYSTEM PRODUCES ACCIDENTS; EMPLOYEES SUSTAIN INJURIES.

COVENANT #6 - TO INCREASE THE 'BOTTOM LINE', MANAGERS MUST EFFECTIVELY MANAGE THE MIDDLE LINES, OF WHICH, THE COST OF L.O. S. S. (LACK OF SAFETY STRATEGY) IS SIGNIFICANT.

COVENANT #7 - SAFETY PERFORMANCE IS A CLEAR AND REFLECTIVE MEASURE OF AN ORGANIZATION'S LEADERSHIP (VALUES), MANAGEMENT (COMPETENCIES), AND OPERATIONAL PROCESSES (SYSTEMS).

COVENANT #8 – A 'CORE TRUTH' IS DEEPLY EMBEDDED WITHIN THE VALUE SYSTEMS OF ORGANIZATIONS THAT REPEATEDLY GENERATE POOR SAFETY PERFORMANCE, AND IT IS: 'PEOPLE DON'T COUNT!'

COVENANT #9 - ACHIEVING SAFETY EXCELLENCE IS REQUISITE UPON MEASURING AND MANAGING THE RIGHT THINGS.

COVENANT #10 - SAFETY MUST NEVER BE THE RESPONSIBILITY OF A 'STAFF' FUNCTION; SAFETY MUST ALWAYS BE THE OBLIGATION OF 'LINE' MANAGERS.

COVENANT #11 – DISCIPLINE, IN ANY FORM AND BY ANY NAME, DOES NOT INCREASE THE LEVEL OF SAFE BEHAVIOR IN A WORKPLACE.

COVENANT #12 – EXCELLENCE IS ATTAINABLE…NOW! ALL OBSTACLES ARE SELF IMPOSED 'L.A.M.E.' EXCUSES.

"Forever be a thorn in their sides!"

PASSING A SAFETY EXCELLENCE 'CAT SCAN'

DAN ZAHLIS AND LARRY L. HANSEN

Originally Published February 2005 as' Occupational Health & Safety' Online feature.

In our work with companies striving to become Safety Excellence organizations, we've learned that the biggest impediment to achieving improved performance is an inability to overcome the conventional 'wiz-dumbs' of safety (wrong headed thinking that impedes progress in the right direction) which inhibit organizational change. John Drebinger, organizational consultant says: "You attain the next level of excellence by changing who you are…and you change who you are, by changing what you do." We've learned that there is one additional preface to affecting 'sustainable change', and that is: "You will only change what you do, when you are willing to change what you believe!"

To truly impact organizational performance (and results), safety leaders must change 'what's inside the boxes', the basic beliefs, values, and prevailing assumptions of their organizations. James Champy, author of 'Re-engineering Management' commenting on the limited results generated from corporate 're-engineering' initiatives, observed: "Much of American management doesn't seem willing or equipped to address directly what is often at the real core of operational problems…MINDSET!" To achieve safety excellence, the safety practitioner must facilitate 'transformational' change. By transformational, we don't mean the ordinary run of the mill type of change in conditions and practices, but rather, change of the frame bending, mind altering type…change that impacts values, systems, manager practices, and structure. Safety leaders must guide their organizations through a series of 'Critical Attitude Transitions'…an organizational 'CAT Scan' so to speak. Following are twelve 'Critical Attitude Transitions' and subsequent actions requisite to attaining safety excellence performance:

CAT #1 – A shift from believing 'employees are the problem', to understanding 'process is the solution'.

Slash the Employee Training budget! Stupid people aren't the problem! Double the 'Leadership and Management Development' budgets…enough said! Well, maybe just a word or two more are in order. Training office staff, shipping clerks, and grounds keepers about the hazards of a confined space, fitting a respirator, or how to lock-out process equipment wastes their time, and your money…as do most of the other eleven pre-scheduled compliance topics, which have little bearing on most people's work…yet are obligated to get 'a check' on the corporate audit score.

155

CAT #2 – A shift from blaming the victims, to holding the guilty accountable.

Discontinue Supervisory Accident Investigations...they rarely identify the 'root cause' of organizational accidents--unless, of course you're willing to put them behind one-way mirrors, bring in a senior manager line-up, and grant them full immunity. They're supervisors…they're not stupid!

CAT #3 – A shift from 'SAFE' as an outcome of a program, to SAFE as an outcome of management systems.

Stick your nose EVERYWHERE it belongs. Encroach upon the 'Turfs' of other functions (sucking out redundancy with a straw), create discomfort with your insurance carrier and brokers (by demanding they do something for those commissions), spend money from one budget account to cover the legitimate needs of another (by fixing problems) and, be willing to sacrifice the most sacred cows and long standing bureaucracies of the organization. If your CEO fires you, CONGRATULATIONS! - It worked.

CAT #4 – A shift from 'Safety' positioned as a staff function, to 'Safety' integrated into the line management operations.

Re-structure your organization. Require that 'shared ownership' replace 'forced accountability'. Build unified business systems (function to function), and collaborative processes (line and staff) NOT functional departments! We all may not be in the same boat…but we are all in the same ocean. Imagine washing dishes at home. Does anyone have a children's dishwashing department, a husband's dishwashing department and a wife's dishwashing department? …OR do we just have one process, with one set of tools and equipment to do one task? None of us does anything so complex at the task level that it requires a damned department…or silo…or island…or smoke stack—I think you get the gist.

CAT #5 – A shift from 'managing' by rules, to 'leading' by values.

Eliminate 'Rules Trolls' and the folly they produce. Rules are made to address 5% of the people (who don't follow them), and they alienate the other 95% (who don't need them). Replace rules with 'values based' process guidelines that delineate systematic methods to be taken by 'PEOPLE' to reduce risk. The phrase "Thou Shalt" shall be reserved for a single purpose – "Thou Shalt refrain from using the phrase Thou Shalt!"

CAT #6 – A shift from 'tolerating excuses' to 'obligating performance'.

Eliminate, now and forever, the word 'ACCIDENT' from the corporate vocabulary. The term 'Accident' is too commonly perceived (and used) by managers as 'a fortuitous, unintended, unexpected, 'S-happens' event, a/k/a- an excuse. Replace it with the word 'incident or operational error', i.e., a foreseeable, predictable and very 'manageable' event manifesting from a series of operational oversights. Now that all the excuses have been eliminated...hold managers accountable for improving their process and minimizing operational error.

CAT #7 – A shift from oppressing (via policy, practice and budget) employee involvement to empowering employee innovation.

Grant all employees a 'no approvals required' purchase authority. Five hundred dollars would be good, a thousand much better! Duct tape and cardboard are nice accouterments for the shipping dock, but far too much of it is used to 'retro-fix' production processes and improve workstations. If employees can outperform the brightest ergonomists with duct tape and cardboard, imagine what they can do with a 'no red tape' spending authority.

Note: Allow all employees to trade their authority amongst one another in order to address needs of higher cost and invite the management team to attend an ongoing lesson in teamwork.

CAT #8 – A shift from encouraging adversarial conflict, to enabling cooperative partnerships.

De-lawyer your business. Run an Accounts Payable printout for all expenses flagged as legal services. If purchasing doesn't categorize expenses this way – force them to start TODAY! Lawyers propagate costs. For every legal dollar spent, there will be more brokered to related service providers, and more yet expended on untracked conflict between the organization, and its employees, service providers, vendors and suppliers.

CAT #9 – A shift from Safety 'by chance', to Safety 'by design'.

Donate the BINGO game and other counter-productive games of chance to the local retirement home! Then design meaningful programs that incentivize and reward people based on desired behavior change, and achievement of goaled and quantifiable activities, which are designed to produce better results…not the fortuitous chance of someone holding a card with a lucky number …BINGO!

CAT #10 – A shift from measurements that 'drive down (reported) numbers', to metrics that 'drive down costs'.

Goal a 300% increase in your 'Recordable Incident Rate' for the next calendar year. Don't cause more injuries, demand more reporting! ...and then ignore the OSHA classification and focus on treating the injured person with respect while aggressively trending, analyzing and preventing the root causes of the incidents being reported. In fact, issue an incentive award for every incident reported, irrespective of severity (see item 9 above). And, by the way, goal a 30% reduction in workers' compensation costs during the same calendar year...you'll beat it by a mile!

CAT # 11 – A shift from sub-optimizing performance via discipline to optimizing productivity via reinforcement.

Stop attempting to 'discipline' people into high performance. 'Whack a Mole' sometimes produces winners at carnivals and State Fairs...but is always sub-optimizing in the workplace. People strive to achieve higher levels of performance when they 'wanna'-- are motivated to do so. When they are told they 'gotta...or else', they devote the greater part of their energy and efforts to getting even! And, they generally will, without you're having the slightest clue. Ever wonder why it takes the incumbent practitioner four months to locate their predecessor's work?

CAT #12. A shift from 'financing loss' into the future, to 'eliminating loss' in the present.

Attack your Risk Finance Allocation System. Demand access to your company's risk finance allocation system. Challenge any/all use of insurance methodologies employed to allocate all costs of risk. Insurance is designed to spread risk, and finance losses into the future...not provoke action on loss causes today. In doing so, it creates a false sense of security, raises ultimate cost, and spreads acceptance and apathy amongst the proactive.

Of course, for those that can't pass the 'SCAN'...there are always alternatives. If an organization, over the long run, is unwilling to accept these 'mind-shifts of excellence', or are unprepared to implement the transformational changes they require, there are still a number of options available. Among the more common are:

- Increase the annual Workers' Compensation Premium budget.

- Beef-up orientation and training expense to cover turnover.

- Add staff to the claim administration function.

- Increase the Personnel Department's recruitment budget.

- Hire more lawyers (my cousin is looking for work).

- Set higher production quotes to offset loss costs, and

- Lower the bar on profit margin projections! (Just spin the annual report).

DECISIONS, DECISIONS, DECISIONS…Someone's got to make them!

A UNIVERSAL MODEL FOR SAFETY X-CELLENCE.

SECTION 7

Leadership

"Leadership is the art of accomplishing more than the science of management says is possible."

- Colin Powell

"I don't need commitment from my leaders...
I need my leaders to LEAD!"

~ Mike Taubitz, General Motors

The Universal Model of Safety Excellence

Safety Leadership

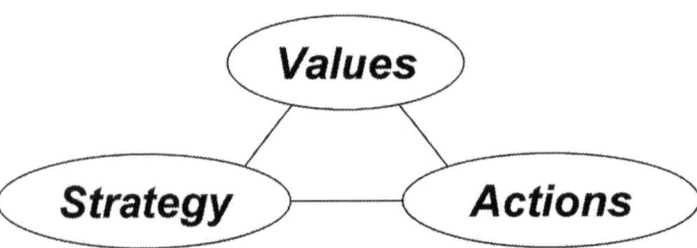

A QUESTION OF EXCELLENCE:

LEADERSHIP

Do our leadership decisions and actions consistently reinforce our values, and demonstrate our commitment to safe operations?

BEYOND COMMITMENT

LARRY L. HANSEN

Originally published in the September 1993 issue of' Occupational Hazards' magazine.

In his recent book, Future Edge, Joel Barker, noted futurist, challenges today's manager to "discover the new paradigms of success." His message is clear: the future will belong to those willing to question current ways, abandon existing mindsets, and break the rules that bind us to the past or hold us in the present.

In most businesses, workplace safety is administered under paradigms of the past – paradigms, which fail to link accident causation to the management systems that ultimately determine all organizational outcomes. Unlike the quality function, safety has not pursued modern management principles to create a 'future orientation."

Tom Peters insightfully identifies the phenomena of a management enlightenment as "… a blinding flash of the obvious!" In safety, this phenomenon exists concerning the linkage of accident causes with their true organizational sources. Most managers would agree that management: designs products, selects and places people, specifies materials, sets policies, develops procedures, acquires machinery, plans schedules, controls the work environment, and shapes the corporate culture.

Yet, despite these clear and accepted management responsibilities, line managers continue to place the cause of workplace accidents on "those careless employees!" This conclusion is wrong! Employees sustain injuries. Accidents occur because of the system and the system is solely the responsibility of management.

Historically, those charged with safety responsibility have cited a lack of management "commitment' as the predominant reason for limited safety success. Being 'committed' doesn't cause anything to happen! Most executives are as committed to safety as they are to all other organizationally and environmentally correct issues, many of which receive little or no attention.

Commitment is a passive state and doesn't generate 'actions' sufficient enough to change the management values and systems, which harbor the true causes of workplace accidents. An illustration by William Hamilton titled, "Do Profits and Social Responsibility Mix." Published in the book Money Should Be Fun, best describes the current limitations of commitment. This illustration depicts a cocktail party and a hardened senior executive responding to an inquiry from a young female business associate. The caption reads: "Committed to safety? Sure I'm committed to safety. I'm committed to safety and any other damn thing that will sell cars!"

As this illustration clearly expresses, commitment is not enough. What is needed from corporate executives is profound knowledge – an understanding of what's right – and proactive involvement – a willingness to act on what's wrong. These are the true keys to management effectiveness, and that includes safety effectiveness!

An effective safety program needs leadership, which is a quantum leap beyond commitment. Only leadership can successfully guide a corporation through a safety paradigm shift to a desired future state: a safe workplace.

Safety leadership (actually a lack of it) can be found in many organizations. If one hears:

- "If only the union would…"

- "If only supervisors would…"

- "If only the government would…"

- "If only corporate would…"

- "If only our employees would…"

- "If only, if only, if only…"

…It's a pretty good bet that effective leadership is lacking.

Leadership is a power of influence created by values and actions. Leaders refuse to assign personal (or organizational) failures, including accidents, to situations or conditions beyond their direct control (i.e., unions, regulators, lawyers, etc.). Effective leaders have an "internal locus on control.' They fully believe that they have the knowledge, skill and ability to influence the ultimate outcome of their efforts and if their results are not what they desire, they know where to look for the causes and answers.

Harold Geneen, past CEO of International Telephone & Telegraph, describes the critical relationship between 'leadership' and 'safety' in his assessment: "having an effective safety program is no different than having a good tennis serve…It's the follow-through that counts!"

If employees don't know what to do; don't know how to do it; have to contend with obstacles beyond their control; are incapable of doing it but are put there anyway; don't have enough time; don't have the proper equipment; are provided with inadequate tools; think they are doing it right yet get no feedback; and are allowed to continue all of the above, then safety programs fail.

Employees in these same companies fail for one simple reason. Anyone care to venture a guess? (Hint: Ten letters – starts with "L.')

Successful safety efforts require executives who are willing to go beyond 'commitment' to participatory leadership. Success belongs to those who are willing to say: "Hold all my calls, I'm in a safety meeting!" (Anthony Skiff, "Safety Works So Why Don't We Use It," Safe Workplace Magazine, January 1993).

That's involvement, that's leadership; and that would be revolutionary!

WHAT SAFETY EXCELLENCE MANAGERS DO

LARRY L. HANSEN

Article originally published May 2006 Occupational Hazards magazine

One of the most commonly cited reasons (read excuses) for under performance in safety is: 'lack of management commitment'…a common allegation: "They just don't care!"

The truth of the matter is: senior managers care deeply, about many things…most importantly, delivering results to shareholders, meeting stakeholder expectations …and 'keeping their jobs!' And, as extremely intelligent individuals, they clearly recognize that the key to their success (and longevity) is their ability to: deliver products and services: at quota, within specification, under budget, on time…and NOT INJURE ANYONE IN THE PROCESS!

Executives of high performing organizations have a clear understanding of Peter Drucker's contention that: " The first duty of business is to survive, and the guiding principle of business economics is not the maximization of profit, it is the avoidance of loss (Drucker). Consequently, they recognize that 'loss, cost, and expense' (the middle lines) are the only things a manager can truly manipulate (read 'manage'); and by effectively doing so, transform top dollars (revenue) to bottom dollars (margin)…the ultimate objective of every manager!

James Champy, author of 'Re-engineering Management' observes: "Much of American management doesn't seem willing or equipped to address directly what is often the real core of operational problems…MINDSET. And, in safety, it is 'traditional safety mindset' concerning accident causation, and correct strategy that impedes most organizations from attaining safety excellence results.

It keeps Going and Going and Going…

In the 1930's, H.W. Heinrich set the safety discipline on a course it has not yet been able to correct. From his original 'theory of causation', has evolved an embedded managerial belief that workplace accidents are primarily caused by but one thing: 'PDDT' - People Doing Dumb Things! more commonly referred to as 'Unsafe Acts'. Based on this belief, (people are the problem), traditional strategies have evolved to focus on Education (Orientation and Training), Enforcement (Rules and Regulations), and since workers are really 'sly and cunning'…Engineering (Safeguards) to 'idiot-proof' equipment and the processes. But, as Karl Albrecht, Quality Consultant has observed: "You work hard to idiot proof a process, and sure enough, someone goes ahead and develops a smarter idiot!"

Today, in the year of the Dog…not much has changed!

A recent survey of the profession indicates that the majority of US companies continue to employ traditional safety strategies: Training – 81%, Compliance – 74% and Technical controls – 75%. (ISHN) Bottom line; 'what we believe about accident causation' and as a consequence, 'what we do about safety' has changed very little over the past 50 years.

And, results?…they have (and continue) to change rapidly…for the worse!

According to National Council of Compensation Insurers data, although total Workers' Compensation cases filed over the past 10 years have been reduced significantly (over 34%), THE COST OF WORKPLACE ACCIDENTS CONTINUES TO ESCALATE, with annual Indemnity and Medical costs increasing 7.4% and 9.0% respectively (NCCI).

No Quick Fixes, BUT…

ACCIDENTS, INJURIES, and their financial consequences–called "L.O.S.S." - Lack Of Safety Strategy (the middle lines), all have a common trigger, (at-risk behavior), and multiple common causes (performance drivers). The Professional Safety May 2000 article "The Architecture of Safety Excellence" identifies these performance drivers as the 'Strategies of Excellence', specifically organizational: Culture (values), Leadership (actions), Structure (relationships) and Process (performance systems). These variables interact to produce organizational performance (safe or unsafe), and ultimately determine outcomes and results, (profitability and sustainability). Although, there are no 'quick fixes' in business (or safety), companies that forge strong values, lead people, align roles and relationships, and effectively manage human performance can reap 'rapid returns' and dramatic reductions in accidents, injuries, and loss costs!

In an organizational safety context, there are three types/levels of change, each of which addresses a different target, and correspondingly, has an increasingly greater impact on performance and results, these are:

- **Level I: Transitional (minor) Change:** Initiatives to change working conditions and behaviors, a/k/a - Safety Programs.

- **Level II: Transactional (moderate) Change:** Initiatives to change organization (roles), process (systems), and management (practices)…a/k/a Safety Management and,

167

- **Level III: Transformational (major) Change:** Initiatives to change organizational culture (values) and executive (actions), a/k/a - Safety Leadership.

To significantly impact results, an organization must advance beyond Level I traditional thinking in safety, (programs), and target Level II (process), and Level III (culture) change, as these are headwater cause of (and controls over) loss in an organization...these are the 'Excellence Strategies' that drive human behavior.

To achieve high performance, leaders must ultimately face the two core questions of safety excellence:

1.   How many 'want to be' a safety excellence organization? and

2.   How many 'are willing to DO' safety excellence?

What many senior managers openly confess about their role is: "They don't know what to do!" They ask: "What 'strategies' should I develop?" "What tactics should I employ?" and "What 'actions' should I take to improve safety performance and bottom line results in my organization?"

This article responds to these questions with best practice citations that identify what high performance managers 'DO' to establish their values, demonstrate their leadership, align their organizations, communicate their expectations, motivate behavior, measure performance, and reward results in safety. These are the targets of safety excellence managers:

VALUES – Employees perform 'safely' when they believe 'safety' is important to the success of the business.

> *"Understand that what we believe, precedes policy, procedure and practice."*
>
> ~ DePree, Past CEO,
> Herman Miller Co., Inc.

EXCELLENCE MANAGERS 'MANAGE BY VALUES'

- **Saint GoBain** – In order to 'manage by values', you first must have values. Bob Scherer, GM of Saint Gobain's Granville, New York Performance Materials plant has documented 19 personal values that he believes will guide his organization to 'safety success' (and they have). He personally meets with and discusses these core values with each new employee during orientation, and then continually reinforces these by conducting on going 'values affirmation' meetings with various plant departments and employee groups. (Scherer)

- **Bechtel Group, Inc.** Does NOT make safety a 'priority'... said Kevin Berg, Principle VP of Safety for Bechtel, Inc. at the first ASSE Symposium on World Class Safety. Safety had always been espoused as a priority at Bechtel, but in recent years it has been changed from a priority to a corporate 'Value'. "When you prioritize something, that means it's not always going to be at the top of your list. A core value is woven into everything you do, every business decision your make." At Bechtel, Involvement, not commitment, drives excellence in safety. (ASSE)

- **Alcoa** – At Alcoa, 'True North' refers to its core values, and in safety that means more than zero injuries. The 'True North' concept, means forward thinking -- "thinking as far ahead as you can think, and then thinking further." Paul O'Neill, past CEO believed: "The absence of accidents does not in any way assure the presence of safe." According to William O'Rourke, VP of Safety, his job is to find out how the company can improve further. "We have to go past zero," he says, "We have to send employees home healthier than when they came into work." (Atkinson and Smith)

LEADERSHIP – Employees perform 'safely' when leader actions demonstrate that 'safety' is important.

> *"If you haven't got any skin in the game; you're not in the game."*

169

-Aussie Football Saying
(Thanks Kelvin Blackney)

EXCELLENCE MANAGERS ARE 'SAFETY VIPS'…VISIBLE – INVOLVED – PARTICIPATIVE!

- **Chevron Chemical** – At Chevron, 'how' a manager gets results, counts as much as the results they get. At a Behavioral Safety Conference in Las Vegas, Jack Beers, Managing Consultant identified 12 specific leadership behaviors, which the company believes supports performance excellence. Employees rate their immediate supervisor on these key leadership behaviors via a confidential 1-800 call in number, and these ratings form part of the supervisor's overall performance rating. (Beers)

- **NYS Power Authority.** – At the NYS Power Authority, the President shows up at safety meetings…all of them! Eugene Zeltman, President of NYPA, observes: " We recognize that safety requires a concerted effort by everyone, from union and nonunion workers to management…and most of all me." Zeltman attends all the quarterly corporate safety committee meetings, which can last as long as two days. In fact, he has not missed one of those quarterly meetings since coming on board as CEO in 1997. "Word has filtered down to employees that if the president attends the meetings, then safety must be important," says Noel P. DesChamps, Director of Power Generation Support Services at NYPA. (Atkinson and Smith)

ORGANIZATION – Employees perform 'safely' when roles, responsibilities, and relationships are well defined and aligned in an organization.

"Every organization is uniquely designed to exactly produce the results it achieves"

-Stephen Covey

EXCELLENCE MANAGERS INTEGRATE 'SAFETY' INTO THE BUSINESS PROCESS… SAFE IS 'HOW WORK IS DONE', NOT A PROGRAM.

- DuPont - DuPont credits its safety success to a philosophy that makes line management - not their 750 environmental, health and safety professionals - personally accountable and responsible for safety, health and environmental. (Smith a)

Line managers are responsible for the incident investigation process, for making employees clear on what is expected in terms of safety performance, for conducting safety training and for integrating safety, health and environmental expectations into the fabric of how work is carried out daily.

- Delta Airlines - Safety is incorporated into every single job description and leadership performance evaluation at Delta, representing the company's requirement for safety in every job function. "Employee involvement in any safety process is critical to achieving success in Delta's operation," says James E. Swartz, director, Corporate Safety. "Safety is a fundamental element in the Competency Modeling process, which describes the characteristics, skills and abilities of people that are related to success. (Smith b)

PROCESS – Safety excellence is but one outcome of an organization's core performance management process, key components of which are: Communication, Measurement and Consequence Delivery systems.

"People, however different, when placed in the same system, tend to produce similar results."

- Peter Senge

COMMUNICATIONS – Employees perform 'safely' when communication systems and practices establish clear expectations, provide timely information, and allow undistorted feedback on 'safety.'

"Employees are in the best position to prevent loss, but they need open channels to share their ideas."

-Tillinghast
Towers Perrin WC Study

EXCELLENCE MANAGERS COMMUNICATE EFFECTIVELY.

- **Steelcase Corp.** – of Grand Rapids MI, sends a very clear message that safety is not a 'competitive advantage' for their organization…it is a 'collaborative advantage'. Steelcase believes that health and safety are so important to their organization that they dedicate a full day each year to a safety conference for 'all managers'. In 2004, realizing that additional 'teaching and learning' opportunities existed, Steelcase expanded invitation to include their vendors, service providers, members of the local business community, and their competitors--but 'no cameras' please. Steelcase knows safety is a 'win – win' proposition for all.

- **Hemerich & Payne** - 'H&P' a drilling company in Oklahoma City, doesn't wait for 'near misses' to react to potential safety problems. Warren Hubler, VP Safety of HP has implemented a corporate wide 'Good Catch' program that incentivizes employees for identifying and communicating 'situations' that can generate potential injuries and incidents. These situations are analyzed, corrected, written up and distributed throughout the organization for learning purposes. Talk about the payback of effective communications; the process 'caught and prevented' a very real potential rig loss valued at a 'one half million dollars from occurring!

MEASUREMENT - Employees perform 'safely' when the metrics upon which they are measured make 'safety' an important measure of their performance.

> *"What gets measured gets done, however what gets done,*
> *may defeat the purpose of what is measured."*

> - Dan Zahlis, President
> Active Agenda, Inc.

EXCELLENCE MANAGERS MEASURE THE RIGHT THINGS, AND PAY ATTENTION TO WHAT REALLY COUNTS.

- **Foamex Inc.** – Foamex Safety Director, John McLaverty, established a Safety Measurement Improvement Team (SMIT) tasked with combining 'leading edge' safety (activity) indicators with 'lagging' (results) incidents measures to create a composite metric which would incentivize facility managers to accomplish safety goals, and to provide a scorecard of their efforts. At Foamex, a good safety performance is achieved when managers…'do more safety' and 'have fewer incidents'. At Foamex 'SMIT Happens'!

- **MeadWestvaco** – "I hate to use a reactive measures like the total case incident rate as the primary measure of the safety process," says Finn Schefstad VP Safety Management.

  Instead, safety excellence process reviews are performed at business units that focus on proactive and preventive safety measures. The objective of the reviews are to: determine where a site is relative to implementing the safety excellence process; evaluate the level of understanding and application of the principles associated with the program's key elements; identify opportunities for performance improvement; and leave the business unit with a blueprint that will move them to the next level of safety excellence and produce sustainable results. (Smith a)

- Active Agenda, Inc. - Dan Zahlis (now President of Active Agenda, Inc.) in a past life as the Western Region Risk Manager for the Häagen-Daz Company, developed what he refers to as 'the ultimate metric of safety'. Faced with corporate pressures to improve Workers' Comp results in the highly volatile California environment, Dan took radical steps to address the real cost driver that plagued operations… he replaced OSHA Incident rates with 'Truth'!

Instead of measuring 'lower recordable rates' (which created an atmosphere of fear and underreporting), he incentivized, measured, and rewarded 'reporting of ALL incidents'—yes folks ALL: Near Hits, First Aid, Medical Only, and Lost Time & Restricted.

The metric used to track performance was 'Total Cost per Incident' – calculated as: "Total injury costs - divided by - the total of ALL incidents. By design of this metric, the only way the operation could truly improve performance was to either drive down injury costs (by managing people better), or drive up incidents (to learn more about risks previously not reported) in the operation. Encouraging 'Truth', built trust, removed the 'veil of fear' that had discouraged past incident reporting, and dramatically reduced the Division's total Injury Costs. (Zahlis)

Note: *The Active Agenda is an open source' FREE' Risk Management technology project. Visit www. ActiveAgenda.net to learn more about this powerful Risk Management Operating system.*

CONSEQUENCE DELIVERY – Employees perform 'safely' when significant consequences (positive & negative) are attached to their 'safety' performance.

> *"You simply can not manage yourself out of problems you behave yourself into."*

> - Stephen Covey

EXCELLENCE MANAGERS MANAGE BY…(AND ARE MANAGED BY) PERFORMANCE CONSEQUENCES.

- **Potlatch Corp.** – At the Potlatch Plywood Mill in St. Maries Quebec, safety performance 'rates high' in importance…for a good reason. A supervisor's overall rating can be no higher than their rating for safety performance, regardless of how well they do in meeting other goals. This practice helped this facility reduce their LT injury rate by 76% and number of lost workdays by 90%.

- **Alcoa** - In the company's 2001 annual report, CEO Alain Belda noted that Alcoa had intensified efforts to raise safety performance at locations, and had undertaken a company-wide effort to eliminate fatalities. As part of that effort, managers of facilities that are under performing in safety should expect a phone call from a member of upper management, as should the managers of facilities where injuries have been reported. Sometimes, he adds, calls go out to facilities when upper management learns safety has improved. "They talk about what the facility is doing well, and what needs improvement."

MOTIVATION – Employees perform 'safely', when they are 'recognized and rewarded' for their performance in 'safety.'

"If you talk about change, but don't change the recognition and reward systems, nothing changes."

-Paul Allaire, Past CEO,
Xerox Corp.

EXCELLENCE MANAGERS USE POSITIVE REINFORCEMENT, RECOGNITION AND REWARDS TO MOVE 'SAFETY' FROM 'GOTTA TO WANNA'.

- **Georgia Pacific Corp.**–GP facilities regularly host what the company calls "Revival Tent Meetings," which are designed to re-energize employees about safety. Facility employees are recognized for safety improvements they initiated and often are asked to speak to the group. Employees from other facilities attend the meetings to share best practices. Lunch or dinner is served outdoors in a large tent, and plant managers often serve as the 'chefs' for the meal. (Atkinson and Smith)

- **Bronson Healthcare** - Asks all managers to write 12 thank-you notes per quarter, and to show them to their own managers as proof that they were indeed recognizing their employees. Additionally, human resources does random spot-checks on managers, asking to see copies of thank-you notes, and if a manager doesn't have them, he or she is asked to schedule a 'little talk' with the senior leader of the group. They've never had to schedule more than one talk before managers quickly got the message that the organization was serious about this activity. (Nelson)

Jerry Garcia inspired 'deadheads' around the globe with two unique contributions…great music, and his conviction that: "Someone has to do something, and it's just incredibly pathetic that it has to be us." An unfortunate reality concerning safety in business is that far too many managers follow Coderre's Law of Least Resistance: "Given the opportunity to do nothing, most will."

This Law does not apply in Safety Excellence organizations!

REFERENCES

American Society of Safety Engineers, Symposium of World-Class Safety, New Orleans, March 2004.

Atkinson, William, and Smith, Sandy, America's Safest Companies – 2002, *Occupational Hazards Magazine*, Penton Publishing, Sept. 2002.

Beers, Jack, "Achieving Important Business Objectives at Chevron with Behavioral Analysis", Presentation at Behavioral Safety conference, Las Vegas, 1999

Coderre, Paul, Personal Communication.

Drucker, Peter, F. "Technology, Management, and Society", New York: Harper and Row Publishers, 1972.

Hansen, Larry L., "ROC Your Organization: 52 Ways to Instigate Radical Organizational Change for Safety Excellence". L2H Speaking of Safety, Inc. Baldwinsville, NY, 2001.

Hansen, Larry L., "The Architecture of Safety Excellence", *Professional Safety*, May 2000.

Industrial Safety and Hygiene (ISHN) 'White Paper' – *Survey of the Profession,* 1999.

National Council of Compensation Insurers, *State of the Line Report* (SOL), NCCI website, 2004

Nelson, Bob (The Guru of Thank You) Website Resources and posted articles (www.nelson-motivation.com).

Scherer, Bob, Saint GoBain Performance Materials, Presentation at Business Excellence Conference October 2005..

Smith, Sandy (a), America's Safest Companies – 2003, *Occupational Hazards Magazine*, Penton Publishing, Sept. 2003.

Smith, Sandy (b), America's Safest Companies Share a Passion for Safety – 2004, *Occupational Hazards Magazine,* Penton Publishing, October 26, 2005.

Zahlis, Daniel F, "CAUTION: Beware OSHA Statistics", *Professional Safety,* December 1995.

Zahlis, Daniel F, and Hansen, Larry L., "Beware the DISCONNECT', *Professional Safety*, November, 2005.

LEADERSHIP – THE DRIVER FOR SAFETY AND HEALTH

OSHA SAFETY AND HEALTH MANAGEMENT SYSTEMS eTOOL – Module 4 - Leadership

A major 1979 NIOSH study concluded that management commitment to safety is the major controlling influence in obtaining success, and overall, maximally effective safety programs in industry will depend on those practices that can successfully deal with people variables.

Safety professional Larry Hansen differentiates the passive management role from leadership when he says safety leadership [is] where executives exhibit 'profound knowledge' — an understanding of what's right — and proactive involvement — a willingness to act on what's wrong.

Leadership is making organizational safety expectations clear, supporting safety financially, being present when key safety issues are decided, being positive about and supportive of others' safety efforts, creating and insisting on a caring company culture. It is, in fact, the single 'overwhelming' factor in achieving an effective safety and health program. Without it, accidents abound.

Above all else, leadership is a constant demonstration by the key managers in an organization, that safety and health are a critical element of daily operations. Here are the kinds of things managers do to show leadership:

- Chair the plant safety committee

- Hold subordinates responsible for the costs associated with accidents.

- Have the safety function, where assigned , report to him/her.

- Have a Board of Director's Safety and Health committee.

- Hold a monthly plant-wide safety meeting where he/she takes questions and addresses safety issues.

- Have any loss incidents reported directly to him/her at the time of occurrence.

- Make sure that organizational safety expectations are absolutely clear by asking every member of the organization about them.

- Be present, and supportive, whenever key safety issues are decided... show they are as important as key product and quality decisions.

- Spend daily time on the plant floor asking people about safety and observing and commenting on issues.

- Start every meeting with a discussion of safety.

- Require a formal safety and health plan from every manager and hold them accountable for results.

- Deliver the safety vision in person to every work team (rather than sending it out in memo).

- Let people see him/her picking up dropped items, moving obstructions, helping out for safety every day.

- Make it clear that any accident is unacceptable and ask hard questions about every one so people know he/she is really serious about having no accidents.

- Act every day in a way that makes it clear he/she knows that everyone is watching to see if safety is really a key value.

- Empower every employee to do what's right for safety...and support and encourage them when they make a mistake.

- Try progressive approaches which fit into the company business strategy/and workplace culture.

- Personally attend all safety training.

- Know the facility safety rules...and never violate any one for any reason...and challenge any one who does.

These examples come from site visits, benchmarking, and trade media reports. They all work somewhere; but few will work everywhere. Success depends on the manager, the company culture, and the nature of the organization. Pick those which will fit.

A UNIVERSAL MODEL FOR SAFETY X-CELLENCE.

SECTION 8

Integration & Collaboration

"There must be not a balance of power, but a community of power; not organized rivalries, but an organized common peace."

-Woodrow Wilson

The Universal Model of Safety Excellence

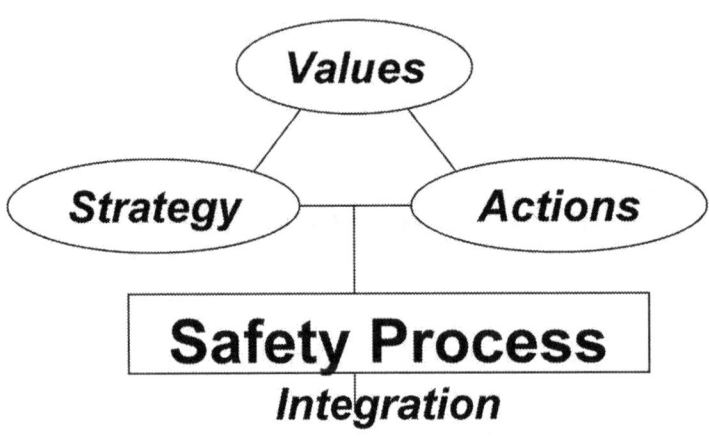

Safety Leadership

Values

Strategy   *Actions*

Safety Process

Integration

A QUESTION OF EXCELLENCE:

INTEGRATION
Have we designed and aligned our organization so that teaming is an integral part of operations?

INTEGRATION

To adequately solve complex organizational problems, managers must overcome 'blind spots' created by narrow functional perspectives. In this complex world, this requires technology.

Technology can help overcome 'Specialty Stupidity': "The inability to see broad organizational issues and ultimate truths due to narrow functional perspectives, a/k/a – 'Blind Spots'.

Blind spots are abundant in business organizations. They are generally caused by specialization, functionalization, and/or competing agendas. When people spend their entire careers promoting a specific function (i.e., sales, production, finance, human resources, engineering . . . or SAFETY), they develop a great deal of expertise in that specific area. But often times -- most times -- problems in an organization have complex multiple causes, causes that go beyond a single function and need to be viewed through a wider organizational lens. Cause and effect is a myth. Multiple causes and multiple effects are reality.

This is probably best exemplified by my experience as a watermelon farmer.

"I decided to quit my job in the insurance industry and move to North Carolina to grow watermelons for a living. I thought I went about it the right way. I bought some land, cleared it, plowed it, planted it, watered it, fertilized it, and seeded it. Later that summer, I harvested it and transported the crop north to sell it. I thought I had things figured right, but I ended up losing money. Many organizations today operate at a loss, or close to it. What do you do? What changes must an organization make to be successful? I can tell you there's no shortage of advice -- here are the options I got:

- My Sales Manager said we must sell more, and to do so we had to increase plantings.

- My Quality Control director was convinced we needed hybrid seeds.

- My Transportation Director convinced me we needed more efficient trucks.

- My Director of Market Research questioned my target market, and suggested cantaloupes!

- My Finance Director advised we needed to leverage our debt.

- My R&D manager was convinced that the answer was in growing 'square melons' (better utilization of truck space).

- The Human Resources Director assured me the problem was team training.

- The Production Manager was demanding new equipment.

- The field workers complained about 'low wages.'

- The EPA wanted me to stop using chemical fertilizers.

- The Agricultural Agent suggested we diversify with corn to replenish the soil.

- My safety director recommended we buy 'safer tractors;' and

- Lastly, my Accountant explained that you can't grow watermelons for $1.50 and sell them for $1.25 each -- at least not for long!

The Bottom Line: If you really want to understand a problem, you can't look at it from an individual, functional, or narrow specialty perspective. You must bring together the composite experience of the organization to look at common issues across a broad spectrum. There is no one 'right answer.'

Source Credit: Another unknown, but very insightful author.

THE HIDDEN AGENDA

DAN ZAHLIS

Originally Published as May 1998 cover/feature of' Professional Safety' magazine

Priority is better achieved through the ability to capitalize on similarities than the ability to highlight differences. It is time to focus on what safety has in common with other functional objectives rather than demanding respect for differences and allowing leadership to 'agree.'

"People First." "Safety First." "We begin all staff meetings with a discussion of LTAs." "Our strategic imperatives, in order of importance: safety, quality, efficiency, cost." These are mere slogans. Many companies claim that safety is a top priority and believe that they fulfill this commitment simply by claiming it to be. One firm's mission statement incorporates this admirable claim: "Safety is the number one priority in our total quality commitment."

In such an environment, those charged with moving the safety effort forward are forced to compete for resources with functional objectives that allegedly owe reverence to the safety of the workforce. The safety function typically fights diligently and fails honorably. The honor comes in knowing that all company slogans, future meetings and informal discussions with employees will verbally promote the priority of safety and reinforce respect for the safety function.

Allowing artificial slogans and meeting agendas to define reality does more than merely slow the process of improvement. Accepting these common practices validates the inactivity that they instill and ensures stagnation. In other words, it is time to "do so and quit saying so."

How does today's safety professional begin to 'do so' when reciting non-compliance risks, warning of high costs and associated low morale, and providing reminders of the value of human life frequently fail? When safety is often 'talked about' but rarely 'walked about'? When the argument that "we are in the business of manufacturing widgets, which generate revenue, not in the safety business" must be acknowledged?

Perhaps the key is to stop talking about safety per se. Instead, the safety professional must begin to explore those functions that receive attention, then design joint processes, which serve the 'ultimate master' (the employer). S/he must join forces, add value and begin to blend safety principles into the culture of business. Today's safety professional must become a resource to other functions and strive to integrate safety objectives into the existing processes of resource-rich departments, functions and objectives. In other words, the expansive supply of support tools and techniques that constitute safety must be applied across the organization.

Clear parallels exist between safety and quality. Long before it was popular, safety professionals were unleashing the power of employee ingenuity. Review of Deming's 14 Points for Management reveals many basic threads of safety. A firm that understands the need to eliminate redundancy and non-value-added activities has recognized the immediate impact of allowing the safety function to act as a vehicle for organizational development and improvement. This does not mean the company should align safety program implementation with that of an organized quality improvement process. Rather, it should use safety as a tool to instill the principles of total quality.

The following examples illustrate cases in which firms have integrated safety systems into other areas in order to capitalize on similarities and tap into the resource-rich employee population.

ORGANIZATIONAL STRUCTURE AND SUPPORT

Figure 1 illustrates traditional efforts to carmelize the workforce around employee protection. The safety function is a resource position attached to a manager (usually human resources or engineering), who is subordinate to the facility leader. In more and more companies, the safety practitioner is attached directly to the leader (Figure 2). Although this shift appears to be an improvement, unless the role itself is regarded differently, firms are doomed to fall victim to the fallacy that one person can accomplish anything independently. This is especially true since the safety professional is often in the least likely position to command or influence performance through position alone.

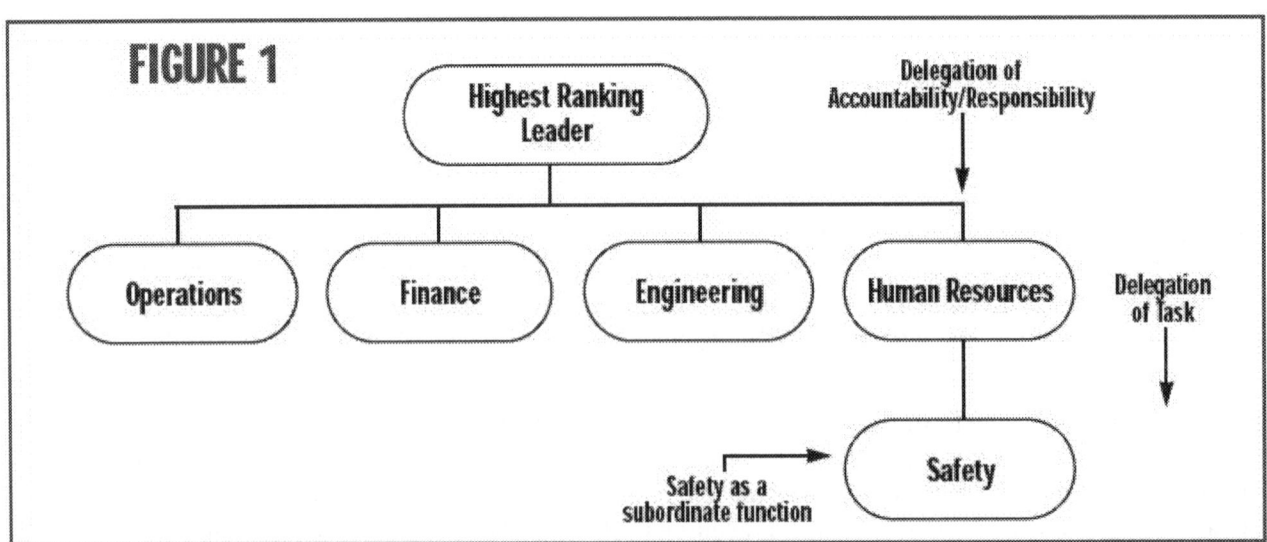

FIGURE 1

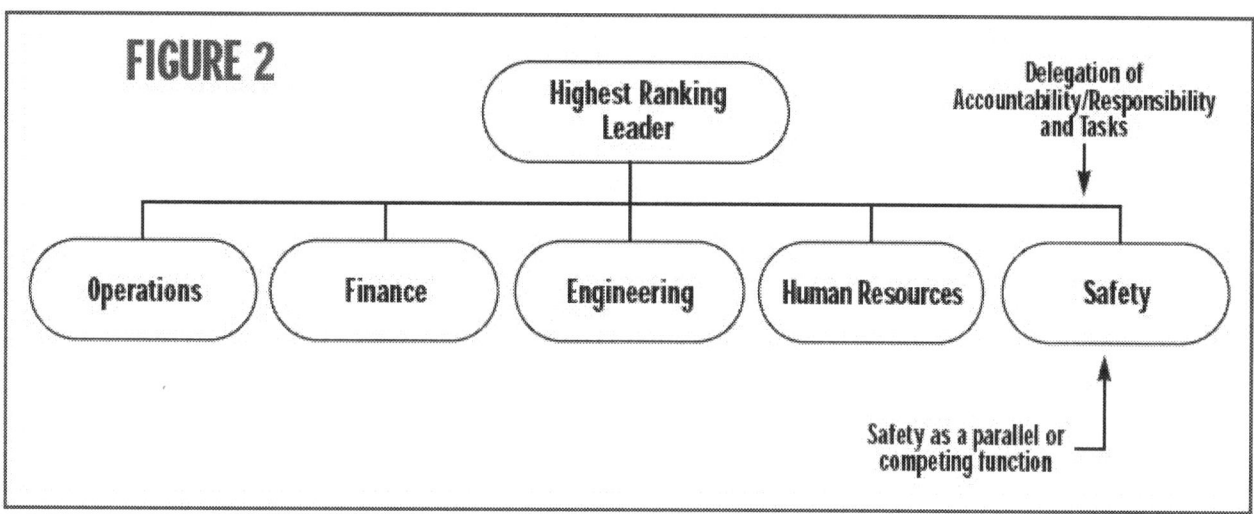

To ensure that the safety practitioner is not viewed as the sole overseer of safe working conditions, a firm must begin to view the highest-level leader as the driver of the effort and the practitioner only as a catalyst and resource (Figure 3). This capitalizes on the practitioner's relative freedom to travel within an organization, without inherent territorial suspicions, while leaving program directorship in its appropriate location – with the leader.

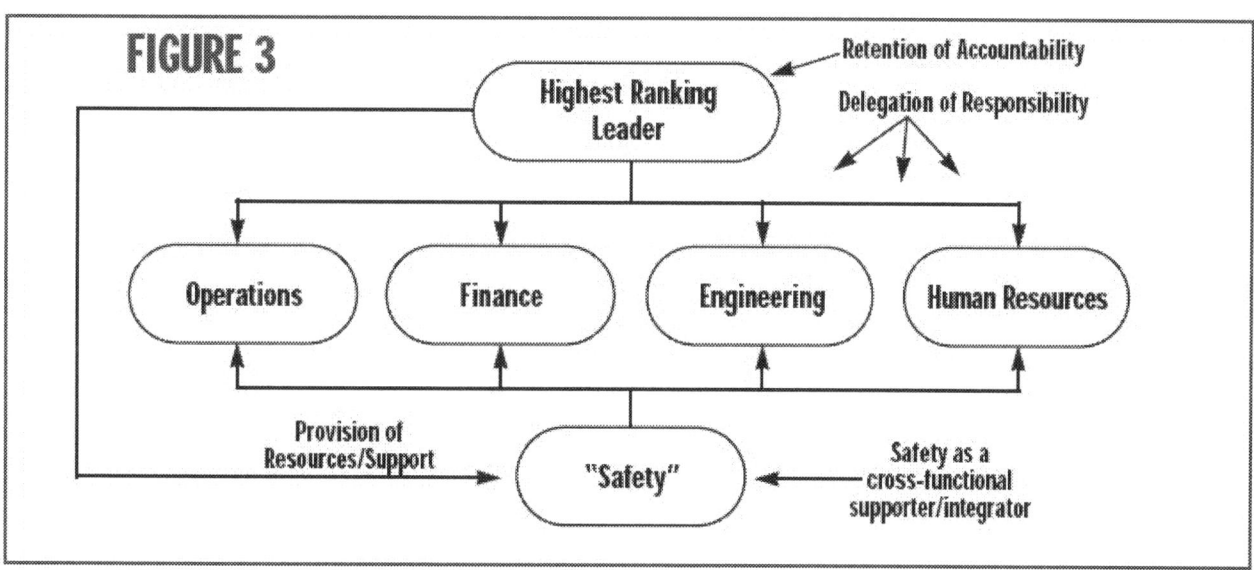

Once the highest ranking leader is recognized as the driver and the safety practitioner as a resource, program responsibilities can be delegated throughout mid management and supervisory ranks – and, in some cases, to line employees (Figure 4). This distribution should be based on individual abilities, with less weight applied to the actual organizational function of program leaders.

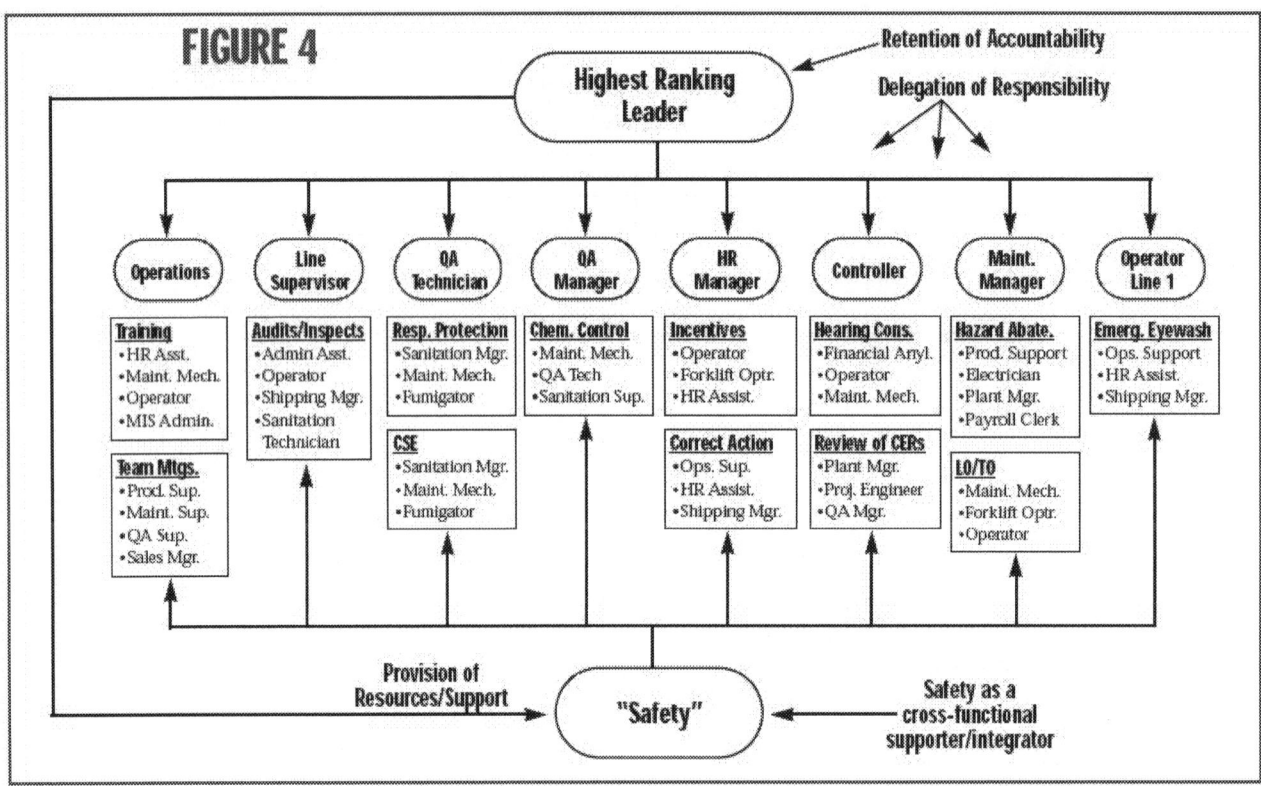

FIGURE 4

Once the structure is identified, program leaders should solicit a cross-functional volunteer team of employees most affected by the respective programs. (Although direct program involvement is not a prerequisite, it does help.) This small team-based approach eliminates reliance on larger safety committees, expands employee involvement and enables teams to make rapid, lasting contributions.

Individual program leaders constitute the steering committee, and they report directly to the leader (regardless of the formal organizational hierarchy). Teams are also encouraged to identify opportunities to integrate additional functional objectives and expand their influence.

One obstacle for team leaders will be their lack of skill in organizing teams, identifying objectives, ensuring cross-functional communication, and facilitating meetings and problem solving. The good news: Team leaders of this "organization within an organization" become the target for leadership development. The better news; Skills to be developed apply across functions and ultimately lead to a stronger business. Some of these skills include giving/receiving feedback; conducting effective meetings; determining root causes; building and maintaining trust; and encouraging initiative.

In such an environment, the safety practitioner becomes a technical resource, staff development facilitator and coach. The 'safety program' being developed becomes an exercise bike that helps fine-tune leadership and empowerment skills. The structure provides leadership practice via small, manageable objectives and organizations (teams). As a result, the learning is real – and is applied immediately.

STEERING THE EFFORT

An integrated firm sees the value of abolishing the traditional 'safety steering committee,' 'plant safety committee' and 'safety coordinator' roles and utilizing administrative committees. During meetings, safety is one area of discussion. Local leaders can also discuss system progress and system needs from a general administrative perspective and satisfy many needs simultaneously (e.g., safety, product quality). Via such an approach, the leader can examine the effectiveness of standard skills being applied to various functional areas; correct divisions; and identify improvement opportunities.

BEHAVIOR MODIFICATION

Modifying behaviors to support a common direction requires continued reinforcement of desired behaviors. Many firms address the issue of behavior modification - reinforcing and correcting – as separate functions. Consequently, they overlook the common attributes these functions share. Methods used to reinforce safe behaviors and correct unsafe ones are common and can be learned under one strategy applied across functions. By integrating behavior modification techniques, a firm can streamline both the tracking and learning process – and improve the consistency of communication throughout the organization.

REINFORCING BEHAVIOR

Figure 5 (next page) represents an early version of one company's list of rewardable activities. Notice that the list is not limited to safety, industrial relations or quality. Instead, it incorporates several functional objectives, and shows that awards are highlighted in a single catalog. By integrating the types of rewardable activities, this firm has developed a method of illustrating the interdependence of various functions. In addition, employees see the flexibility of the list and are encouraged to add items they feel reinforce the process.

Also notice the absence of rewards attached to the occurrence or non-occurrence of downstream effects (e.g., LTAs, consumer complaints, efficiencies). This clearly demonstrates a preference for activities that develop the organization and reinforce group contribution through individual accomplishment.

CORRECTING BEHAVIOR

A company can enhance behavior modification by improving the techniques of those charged with correcting employee behavior. To accomplish this, the firm can develop a training method that utilizes real-life examples as interactive exercises during training. These exercises should not be limited to one functional area; rather,

187

they should cross functions based on key focus area (e.g., cost, safety, product quality, customer service, efficiency). By integrating leadership techniques, skills learned once can be applied repeatedly.

Traditional efforts to teach line supervisors how to take disciplinary action in response to safety violations simply reinforce the notion that safety is different from other functions and, therefore, requires a different technique. This only further distances safety from the organizational 'big picture' and builds resistance to safety because of the need for additional learning and resources.

TEAM MEETINGS

Safety meetings are generally mandated by OSHA and certain state programs. Many

FIGURE 5

CRITERIA	POINT VALUE
Non-repeat suggestion	1
Suggestion implemented	3
Catching a wrong code date	2
Reporting unsafe condition (product or human)	5
Conducting a monthly self-inspection	4
Obtaining a required certification	5
Serving on the plant ergonomics committee	5
Finding and removing a non-naturally occurring foreign object from product	4
Becoming trained as a program trainer	25
Serving a term on a systems team	10
Specialized training received	20
Zero absences	1st six months 5 2nd six months 5 1 year 25
Team meeting presented	25
Verified report of a suspected fraudulent claim	Employee allowed to select a gift from the catalog.
Reported an unsolicited contact by a workers' comp practitioner (doctor, chiropractor, lawyer, physical therapist, etc.)	25
Best suggestion of the week (determined during staff meetings)	5

firms conduct these meetings to accomplish basic compliance, so they tend to overlook the cross-functional opportunities such gatherings present.

Figure 6 depicts a sign-in sheet utilized by a firm that recognized the value of associating safety meetings with other key strategic imperatives. Each topic listed must be covered during each meeting.

This method of topic integration enhances the likelihood that such meetings will occur because it builds cross-functional support for their occurrence. This also means the safety coordinator need not coax participation. Clearly, the adage that a safety meeting should only cover safety topics in order to avoid weakening the message is outdated and unrealistic.

INSPECTIONS/AUDITS

OSHA mandates routine inspections of the workplace, while agencies such as the U.S. Dept. of Agriculture require self-inspection of plants that process and package food. Quality assurance and sales departments depend on regular inspections to ensure that specifications are met. Operations groups depend on regular process reviews to reduce the likelihood of downtime and product risks. Maintenance relies on inspections to facilitate preventive actions. An integrated firm designs and utilizes common tools to accomplish multiple objectives – including inspections.

Figure 7 represents an abbreviated version of a pre-startup checklist used by a food-processing firm. This list integrates the myriad regulatory

FIGURE 6

On this day, _____, a team meeting was held for the _____ department/line/team. The individuals listed below attended. Each participant was encouraged to interact and ask questions when necessary. The topics covered during this meeting were:

| Safety: |
| Product Quality: |
| Shrink: |
| Line Efficiencies: |
| Labor Relations: |

This meeting (did/did not) result in a recommended modification to an existing practice, system or process. The recommendation is as follows:

Supervisor/Team Leader Signature _____ Date _____

Last Name First (please print)   Signature

1. _____   _____

2. _____   _____

FIGURE 7

Date and Time of Inspection:	__/__/__ AM/PM M D Y Time				Page 1
EQUIPMENT	INSPECTION ELEMENT	OK	NO	ACTION TAKEN	
PRODUCTION	Floor area is free of debris.				
	Pest traps are satisfactory 1,33, 34.				
	Floor drains are covered. Covers in good repair.				
UTILITY AREA	Packaging materials are staged and are secure.				
	Water hose is in good repair and is coiled.				
JONES CARTONER	Feeder guards and caution stickers are in place.				
	Machine is free of dust and residue.				
	All panel lights are operational.				
	Feeder chain is free of excess oil and debris.				
	Floor mats are clean and in place.				
	E stops are functioning properly.				
FEEDING BELTS	Belt alignment is acceptable.				
	Feed belt area is free of product from previous run.				
	Guards and caution stickers are in place.				
SOLBURN FILLER	Rotation is adequate.				
	Filler area is free of mold, slime, debris and odor.				
	Safety nozzle and air hose in place , good repair.				
	Fruit capture tubs are in place beneath the filler.				
	Shaker motor oil gauge reads 45 PSI.				
FLAP TUCKER	Water jets are operating.				
	Belts are in good repair.				
	Fruit capture tray is in place.				
	Caution stickers are affixed to access doors.				
WEIGH CHECKER	Scale is calibrated and accurate.				
	E stops are functioning properly.				
	Weigh checker seat is in place.				

and self-imposed expectations into a common document. As a result, the firm has eliminated redundant physical inspections; illustrated the equal importance of various functions; provided a dynamic tool that incorporates continuous learning; and standardized the technique for conducting inspections.

Initially, this firm's functional leaders resisted integrated inspections, believing that maintenance personnel should gauge maintenance needs; that quality inspectors should monitor product quality; and that safety staff should monitor hazard control. These fears would be well founded within a company in which management operates the facility along functional lines, yet unthinkable in one where management serves the needs of those who truly operate the facility.

HAZARD ABATEMENT

Hazards are hazards, regardless of the target population. Auditing for the presence of hazards under separate functional objectives (e.g., human safety, product safety, property protection) merely promotes functional thinking and generates redundancies. Within such a system, each functional group must track its findings and communicate them (independently) to those charged with implementing corrective action. This process is often achieved via different formats – and on various timetables – which only further confuses those who must decipher the requests. Consequently, it becomes difficult to manage abatement follow-up.

One firm developed a tracking mechanism to streamline its tracking, priority setting and communication methods. This firm incorporated all hazard tracking into one system – with one system administrator. As a result, the company was able to clearly communicate where those wishing to report hazards should go (irrespective of the nature of the hazard). Maintenance and engineering personnel also know where to obtain hazard reports – or where to expect them on a regular basis. In this firm, the leader has a single method of tracking hazard abatement – which is projected to the front of the room during steering meetings – and accountabilities are assigned and tracked in 'real time.'

This firm's system is not designed to eliminate only human safety concerns. It tracks all hazards interdependently and categorizes them for those interested in a particular type of hazard.

For example, if OSHA asks to review hazard abatement, the facility simply accesses the system and sorts by hazards categorized as human safety. If product quality auditors inquire about product safety, only the sort changes – not the administrator, mechanism or manner of follow-through.

What did this organization learn? Differentiating between hazard categories is challenging – and often leads to disagreements because many hazards appear to affect multiple areas, further illustration of organizational interdependence.

JOB HAZARD EVALUATION

Job safety analysis (JSA) is a tool used to identify hazards associated with specific job tasks. The simplicity and effectiveness of this tool has prompted several states to legislate it into administrative program requirements. Yet, JSA is under utilized.

Figure 8 is an edited page from one firm's general job hazard evaluation (JHE), which is a base training requirement for all workers. Employees must also receive training in JHE's relative to their jobs. This firm recognizes the value of documenting task review as it relates to multiple functions that produce hazards.

FIGURE 8

TASK	HAZARD	CONTROL MEASURE	PpPE
ENTERING PLANT	Introduction of hazards to food product	Wash hands prior to walking through plant.	Hairnet Smock
		No smoking, gum chewing, eating food in plant.	
	Collision with forklifts	Remain within the pedestrian walkway along the outside of the yellow berms along wall.	
		Always use pedestrian doors.	
	Contact with moving machinery	Do not place hands into equipment with moving parts.	No loose clothing
	Slippery floors and floor obstacles	Use caution when walking (walk slowly, avoid standing water, do not step over objects - go around).	Non-slip, closed toe shoes
		Remove standing water immediately.	
	Slippery stairs	Always use handrails and walk cautiously on stairs.	
	Excessive noise	Wear hearing protection at all times.	Earplugs
	Contact with low equipment	Wear head protection.	Bump cap
	Unidentified dangerous conditions	Inform a supervisor immediately of dangerous conditions. If you wish to remain anonymous, use suggestion box.	
HANDLING FRUIT	Introduction of hazards to food product	Never touch fruit without appropriate gloves.	Latex gloves
EMERGENCY EVACUATION	Failure to escape a dangerous condition (i.e. fire, ammonia)	Locate supervisors in your immediate area (blue cap).	
		Familiarize yourself with posted evacuation.	
		Proceed with caution in the same direction as employees evacuating the area.	

The employer reviewed all tasks that generated hazards – whether the hazards applied to employees or end-users. In fact the firm created the acronym PpPE (personal and product protective equipment) to reflect recognition of hazards to people and products. The site also uses base task lists, identified via the JHE development process, to create accurate job descriptions and standard operating procedures.

If a firm assigns an employee to list every task performed, why not assign several colleagues from other affected functions to help? The end result will be a single document that all functions continue to modify as learning grows. This integrated approach leads to increased continuity via standardized training materials, leadership accountability, system tracking and modification, and organizational communication.

ADDITIONAL OPPORTUNITIES

The possibilities of program integration are limitless – and today's market conditions demand compliance integration. Other areas of opportunity include training, suggestion programs, chemical control and hazard communication, ergonomics programs, change management, regulatory inspections and contacts, and capital expenditure review programs.

Determining which functions to incorporate depends on the dynamics of the organization's leadership. Operations managers who recognize that leadership truly holds safety as a priority will capture this budget value. Incorporating goals outside the function of safety into a function protected by leadership mandate is essential. In forms that hold the quality assurance function reverent, managers can incorporate safety goals into the quality assurance function.

The key is to acknowledge the similarities between functions and reject the practice of saying one thing and doing another. Many will see the opportunity to implement goals and objectives under the firm's true priority list.

Allowing safety to be mentioned first (as a means of acquiring priority) limits the level of priority to words alone. Embedding safety principles into the operational culture delivers tangible priority to the safety function while expanding the value of safety practitioners. What ultimately matters is what gets done – and whether what gets done actually improves the process of protecting employee and product safety.

So, what is the hidden agenda? Depending on one's perspective, it is either 1) implementation of a participative, sustainable employee protection program or 2) a developmental process which improves communication, participation and leadership skills that lead to a healthier business. The irony? The effectiveness of the safety program depends on the firm's health; in order to accomplish the agenda, safety professionals must solicit the integrated participation of the organization, with an emphasis on leadership. Safety can still be first.

SECTION 9

Human Relationships

"I have never seen an operational problem that was truly about technology, marketing, or manufacturing. It always comes down to . . . people."

~ Jacob Needleman

"With over 50 years of hard evidence at hand, it's hard to 'slough off' the truth. It's all about people."

~ Peter Drucker

The Universal Model of
Safety Excellence

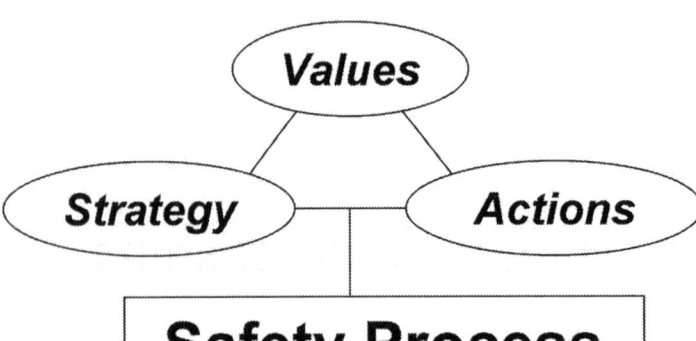

Safety Leadership

Values

Strategy   *Actions*

Safety Process

Integration

Safety Management

Human Relations

A QUESTION OF EXCELLENCE:

RELATIONSHIPS

Do we respect and value our people,
and do our policies and practices
build trusting relationships that bond
employees and the organization?

12 TOTALLY 'UNLEGAL' WAYS...TO SLASH WORKERS' COMPENSATION COSTS

LARRY L. HANSEN

"They're a bunch of freeloaders!" "They don't want to work!" "Sore back, baloney!"

"They're all careless!" "You can't trust any of em!" Sound familiar? It should. These are some popular tunes on today's business "hit parade.' Nat King Cole's classic underscores today's workers' compensation (WC) reality: "Although it's been said many times, many ways . . ." the problem is bad employee attitudes. The management premise that bad employee attitudes drive WC costs continues to be an issue of heated debate. To move beyond fault-fixing, I will concede that 'bad attitudes' are central to the WC problem. Certainly, employee attitudes have changed. 'Generation X' employees have new perspectives, values and expectations of work life. The management side of the equation has also changed. Business has 'RIFed' (Reduction In Force) workers. Rightsizing and downsizing have taken their toll. Consequently, labor/management relationships are now different. Fortune's June 1994 feature, "The New Deal,' characterizes the changing attitudes and loyalties in today's business world. Labor's new values are reflected in the words of one young New Jersey insurance worker. "We're cold and calculating and looking out for ourselves" (O'Reilly 4). Sounds like a line from the movie Wall Street.

THE ORIGIN OF BAD ATTITUDES

Let's cut to the core issue: Who is responsible for these attitudes? As a consultant for 25 years, I have found that all key business issues are reduced to two acid tests: 'cost/benefit' and 'make or buy' decisions. Traditionally, employee attitudes have been a 'make' decision. Firms have invested heavily in 'selecting out' human problems by dedicating an entire corporate function (Personnel) to 'selecting in' right people. Although federal regulations (and the liability courtroom) have established some parameters for this task, results have been fairly successful. Companies do not hire 'bad attitudes' (they are smarter than that). That leaves only one conclusion: Management is highly efficient at 'making' them (obviously not so smart).

Bad attitudes are an issue—but they are not the problem. The true problem is their cause—the reasons they exist and, more specifically, the practices that create them. In other words, corporate values and management practices. Author Jay Michael Crouch concludes, "Although employee attitudes are important, they are irrelevant until management attitudes are addressed." In The Customer Comes Second, Hal Rosenbluth argues that "business earns the bad attitudes of its employees." Both are convinced that the true source of customer satisfaction is employee satisfaction and that management is the source of both. Fortune identifies Chevron, Intel and Reuters Organization as firms that have recognized the need to rethink human resource values and become more proactive in building positive employee attitudes and loyalty. In describing the difficulty of this task, Chevron vice chairman James Sullivan concedes, "Until you try to write about it or talk about it, you don't realize how inept you are" (O'Reilly44). Chevron has implemented aggressive communication and information-sharing processes to explain critical issues and keep employees informed of priorities, directions and accomplishments.

Intel has enhanced its corporate communication efforts via "BUMs" (Business Update Meetings) and 'SLRPs" (Strategic Long-Range Plan) meetings. During these sessions, Intel executives meet with employees to help keep them informed, involved and positively oriented toward corporate objectives. Reuters has restructured its corporate values to have positive impact on employee loyalty and the bottom line. The firm's Celia Berk, who oversees employee development and training, explains, "We decided that if you measure yourself just by financial results, you can't tell if you're creating an opportunity for rivals" (O'Reilly44). Reuters now measures itself based on (in this order): 1) client satisfaction; 2) employee effectiveness and satisfaction; 3) operating efficiency; and 4) contribution to shareholder value (financials).

Procter & Gamble (P&G) Corp. is also a leader in valuing the human side of enterprise. At a 1994 conference, Rick Fulweiler, then corporate safety executive, spoke of his firm's three critical guiding principles. At P&G, people are: •essential to the overall, continued success of the enterprise; entitled to the preservation of their life and health; key to productive, high-performance work systems. Fulweiler shared the convictions of his firm's CEO who, when asked to assess the value of people to P&G, responded: "Procter & Gamble could lose all its plants in a single major catastrophe and conceivably be back in business with restored market share within 10 years. If we were to lose our people, there is no return . . . there is no future . . . it's all over" (Fulweiler). Thanks to such core values, P&G truly 'capitalizes' on its human resources. Progressive organizations recognize that to solve problems, management must take ownership of them. With respect to WC costs, managers must recognize that they have a genuine opportunity to impact results.

The April 1994 issue of the Ohio Monitor examined the successful WC turnaround achieved by American Spring Wire Corp., Bedford Heights, OH. What does James McDonald, vice president of operations, identify as the key driver of the change? "To solve a safety problem," he says, "you need to tackle it the same way you would any other problem. That is, you need to approach it thinking 'I'm to blame, no one else'" (Avers 16). In The Real Heros of Business, Bill Fromm and Len Schlesinger identify the critical impact that 'trust' has on attitudes. "Fifteen years of evidence shows that, on average, only 1 to 1.5 percent of Americans are systematic cheaters." The rest have other reasons. These authors suspect reasons of trust. The trust-attitude-cost correlation is particularly strong in respect to WC. "Accidents don't drive WC costs . . .claims do" (Brooks). Claims (the dollar value of accidents) are largely subjective—a matter of worker perception. An employee's perceived need to lose time, the amount of time lost and the degree of residual disability incurred are all decisions an employee makes based on his/her perception of management's values and fairness.

"Workers are like a mirror, the reflection you see is your own," says Philip Crosby. Employee attitudes are a reaction to management values and actions—a 'make' decision. These attitudes span a broad spectrum, ranging from B.A.D. (Belligerent And Destructive) through a population of average J.O.E.s (Just Ordinary Employees) to the S.A.I.N.T.s (employees who Say All Injuries are Negligible and Temporary). Disability studies indicate that hard-core fraud accounts for only a small percent of all WC cases (i.e., the 'BAD' actors). At the opposite end of this scale are those with a strong work ethic, who would 'work through' an injury and who return to work as soon as possible.

This leaves the average J.O.E.s, who comprise the largest segment of the workforce. These employees react to how they are treated and valued. In other words, some companies take advantage of their employees (rely on them as a resource) while others take advantage of (disregard or exploit) their employees. Employees know the difference and react accordingly. A firm's 'human asset' management skills offer the greatest opportunity to shape attitudes and behaviors and significantly reduce WC costs.

Businesses' poor track record and inadequate practices have spawned an adversarial WC system that is driven by distrust, animosity and 'revenge.' The end result: Inflated costs—the 'legal' solution to WC. The new 'unlegal' way to minimize claims and slash WC costs entails building positive employee attitudes via positive people values.

Following are 12 'totally unlegal' ways to build these attitudes and achieve superior WC results.

1.   TELL IT LIKE IT IS . . .AGAIN AND AGAIN AND AGAIN

The key to success in real estate is said to be "location, location, location!" In all other business, the key is "communication, communication, communication." Often, communication problems are perceived as 'information gaps,' which, in reality, do not exist. When management does not provide timely, accurate information, employees fill the gap with their own—the notorious grapevine. Unfortunately, the grapevine often conveys 'unfacts' and negative information. Or, when the information is correct, management has been 'beaten to the punch,' which reduces its credibility. When labor and management communication networks are not in tune, the result is organizational 'white noise.' Consequently, messages are neutralized."Management broadcasts on WDIK ('What Do I Know?') while employees are listening to WDIC (What Do I Care!). The outcome is chaos" (Coderre). Within the current "no news is good news' business communication paradigm, managers cannot positively impact employee attitudes. Such an approach allows the grapevine to flourish and guarantees that employees will remain uninformed, skeptical and uninspired. To positively impact employee attitudes, managers must share information in a timely, objective manner. Even bad news provided on a timely basis is better than no information. People would rather be 'in the know' than 'in the dark.'

It is 'unlegal' to keep people in the know.

2. DEVELOP THE BIG PICTURE

Traditional top-down organizations isolate employees and cast them in functional roles with jobs that are disconnected and seemingly unimportant. Work that lacks purpose is perceived as unimportant, which leads to low worker esteem and, ultimately, to poor employee attitudes. Managers must ensure that each employee knows the importance of his/her role in the corporate mission. Each employee must understand why his/her contribution is key and how such efforts ultimately impact the end result. It is management's job to frame 'the corporate portrait.'

It is 'unlegal' to bring everyone into focus.

3. ATTEND TO THE SILENT MAJORITY

Listen . . . . Good employees are out there, you just can't hear them. Most managers fall victim to Oterap's (Pareto's backwards) Law—they spend most of their time dealing with problem employees. Managers are drained by low performers (i.e., constantly patching the dike). It is a firm's great performers who deserve the most time and attention. These performers (the silent majority) have the interest, enthusiasm and creativity to propel a corporation forward. Therefore, managers must reverse these time/attention ratios and begin spending more time with those who have answers and less with those who have problems. Answers move companies forward.

Spend time recognizing success—and those who create it.

It is 'unlegal' to give employees reasons to excel.

4. DO UNTO THEM . . . FIRST.

Military strategists know that decisive victories are often won thanks to "pre-emptive strikes'—a first, fast, effective action initiated with the element of surprise. Such tactics catch opponents offguard and neutralize their offensive capabilities, leaving them far more willing to listen and reason. Applying this strategy in the workplace is an effective way to build positive employee attitudes. In describing his deliberations with the Libyan government, Armand Hammer recollected: "When I got to Libya, the prime minister unbuckled his revolver and laid it on his desk. I thought, that's a good way to start negotiations." If positive employee attitudes are the goal, managers must be the initiators of the respect, fairness, dignity and equality that create them. As Jay Willard Marriott Jr. says, "Take care of your employees, and they'll take care of your customers" (and, consequently, you).

It is 'unlegal' to employ pre-emptive attitude strikes!

5.   IF NOT IN THE REFUSE BUSINESS, DON'T OPERATE A DUMP

Although this may seem like common sense, it is not so common in practice. Employees spend a large part of their life in the workplace. As a result, their attitudes are a product of their work environment. How can management expect employees to have positive attitudes and deliver standing-ovation performances if they are cast in squalor? Just as athletes play up (or down) to the level of their competition, employees perform to the level of their leader's expectations. If a manager expects (accepts) poor working conditions, bad employee attitudes—and poor leadership—are the likely result. If management does not care, why should employees?

It is 'unlegal' to clean up your act!

6.   PAY UP—YOU 'OWE' THEM.

Many performance improvement initiatives, including empowerment and self-directed teams, fail because managers will not 'let go.' Lack of trust is perhaps the most self-fulfilling prophecy in labor and management relationships. "Firms which operate with policies that send clear signals that employees can't be trusted, in fact, end up with employees who can't— all the good ones leave" (Fromm). Physicians and attorneys have insightful perspectives of this reality. Brent Lovejoy, an occupational physician in Denver, and Sharon Stiller, a partner in the Underberg & Kessler law firm, Rochester, NY, recognize this critical issue. Both enter the WC system near the end of the process, when emotions and costs are high. Once these emotions are stripped away, both clearly see how focused the core issues truly are—workplace values, people deprived of what they are 'owed.' In other words, trust and respect.

It is 'unlegal' to employ trust and respect.

7.   INVOLVE EMPLOYEES . . . ALL OF THEM.

Within the traditional business paradigm, management has all the answers. The message to employees is, "No thinking allowed." In such an environment, employees learn to follow the rules no matter how stupid. In today's business environment, an employee who is afraid to ask better ask. Questions not asked can kill someone; procrastinating only makes the process more painful. Employee involvement is one of the greatest shapers of positive attitudes and behaviors. The Rome (NY) Cable Co. provides undisputable evidence of this fact in the form of its 'SWISH' (Six Willing Individuals Scoring High) team. Team members Mike Mason, Bruce Allen, John Siniscard, Ron Brodock, Pete Phillips, Mike Potts, along with 'coach' Bill Casanova, are a self-directed work team within the Banbury Dept. for Rome Cable and this team, empowerment, communication and employee involvement have produced significant improvements — including fewer accidents. The department, problematic prior to the company's commitment to teams, has shown dramatic improvement in most performance areas.

When asked what changed their attitudes and, consequently, their performance, the team's message is clear: "We've got minds, we've got ideas, we've got experience . . . we want to be recognized and valued as more than just a pair of hands." Now they are.

In most cases, all management needs to do is ask. Employees are an untapped wealth—an organization's 'undeclared assets.' As Tom Peters says, "There's no limit to what the average person can accomplish if thoroughly involved."

It is 'unlegal' to engage their minds.

8. SEE THE DIFFERENCE IN PEOPLE.

Today, the differences in products and services are often barely perceptible. This leaves people as the only real differentiators of success—the 'personal signature' differences. Excellence in any process is driven by the details, the fine points . . .the personal touches. Unfortunately, in many organizations 'good 'nuf' is just that. No real effort is taken to create excellence by going that extra inch, mile or whatever it takes. In safety, the bar cannot be placed at 'just adequate' —it must continually be raised.

Safety must demand that:

- No job is considered complete until the work area is cleaned.

- No machine is put back into operation until all guards are replaced.

- No short-cut is allowed or rewarded merely because it is quicker.

- No safety and health policy is posted, then ignored.

- No safety goal is established, then excused away.

It is 'unlegal' to 'recognize the personal touch.'

9. CAPITALIZE ON HUMAN ASSETS.

Businesses have a peculiar 'value system': They know (to the dime) what a customer is worth and (to the penny) what an employee costs. Yet, they have no clue what good employees are 'worth.' To change this system, companies need to conduct a 'valuing process' in order to establish the true worth that employees contribute to sales, revenue, profitability and innovation. If firms were to value employee contributions at market cost or the current rate for 'consultants' (which is what employees are), they would be amazed at the value that good employees add. Such an exercise also reveals that, at current wages, most companies are getting a bargain.

It is 'unlegal' to value employees more, and measure them less.

10.   SEEK OUT MAD SCIENTISTS

"Rube Goldberg" was famous for 'rigging' simple solutions to big problems at minimal expense—not pretty, but highly effective. Most organizations simply do not value such solutions—but they should. If the goal is to complete the job at the lowest cost, then perhaps management is too focused on protocol and 'aesthetics.' The paradigm needs to shift from 'what is right' (procedures) to 'what works.' To do so, management must locate the problem-solvers. Evidence of them can be found anywhere—simply search for basic, homemade solutions to difficult process problems. The result may not be pretty but, bottom line, it is likely to be effective. Remove the bureaucracy, approvals, paper trails and organizational constraints, and effective solutions to significant safety problems will emerge.

It is 'unlegal' to recognize good ideas and their originators.

11.   REPLACE RULES WITH THINKING.

If something can be done a better way—even when it is not the 'accepted' way—go with it. Any route forward is a good road to travel. Innovative employees are intolerant of the status-quo, particularly those embedded in rules, procedures and company policies. Smart employees have trouble following dumb rules—and systems. Burger King came close to this philosophy when it declared, "Sometimes you gotta break the rules.' Close . . . but no banana. Employees should not be forced to break rules to be productive. If the rules are wrong, do not apply, are dated or can be improved, they should be changed.

Successful organizations are populated by innovative, creative people who find unconventional solutions to pressing problems. Management can encourage these characteristics by removing the restraints on their creativity. Dana Corp. credits its success to the fact that it eliminated procedure manuals and replaced them with trust. As Voltaire said, "The best don't need rules and the worst won't be helped by them.'

It is 'unlegal' to reward good employees who bend bad rules.

12.   GIVE 'EM LATITUDE, NOT ATTITUDE

Find an employee who irritates his/her boss and chances are that employee is doing 'good stuff.' Effective employees are driven to do right things despite procedures, policies, quotas and other restraints that call for just the opposite. Innovative employees are a challenge to manage because their personal values often exceed those of their supervisor or others within the organization. Many managers perceive their role to be: control, ensure conformance, follow procedures and maintain the status quo. These objectives conflict with creative employees who strive for new, different, better ways.

To truly improve organizational results, get out of their way! Nothing oppresses independent thinkers and innovators more than a hovering manager. Accountability does not drive results, commitment does.

It is 'unlegal' to provide more head room.

"But your Honor, I'm innocent. I didn't do any of these things!"

The verdict: "Guilty as charged!" Case closed! No appeals!

REFERENCES

Avers, Laura. "Strengthening Safety: An Evolutionary Process." *Ohio Monitor.* March/April 1994: 16-22.

Brooks, Dennis. Discussion at Minerva International Conference, Scottsdale, AZ, May 1994.

Coderre, Paul C. Personal communication. July 25, 1994.

Cohen, Harvey H. and Robert J. Cleveland. "Safety Program Practices in Record-Holding Plants." *Professional Safety.* March 1983: 26-33.

Crosby, Philip. Quality Is Free. New York: The Penguin Group, 1980.

Fromm, Bill and Len Schlesinger. The Real Heroes of Business. New York: Currency Doubleday Publishing, 1994.

Fulweiler, Rick. Presentation at Minerva International Conference, Scottsdale, AZ, May 1994.

Herzberg, Frederick. One More Time: How Do You Motivate Employees? Cambridge, *MA Harvard Business Review.*

Mauer, Rick. Caught in the Middle: A Leadership Guide for Partnership in the Workplace. Cambridge, MA: Productivity Press, 1992.

Mitchell, Russell and Michael O'Neal. "Managing By Values: Is Levi Strauss' Approach Visionary or Flaky?" *Business Week.* Aug. 1, 1994: 46-52.

O'Reilly, Brian. "The New Deal: What Companies and Employees Owe One Another." *Fortune.* June 13, 1994: 44-52.

"The Pygmalion Effect in Management."In "Module 1: Setting the Scene for Coaching." Kasset International Inc., 1990.

Stiller, Sharon. Personal correspondence. Dec. 6, 1993. [Author's note: Credit for Articles title goes to Ms. Stiller.] Rosenbluth, Hal and Diane McFerrin Peters. The Customer Comes Second. New York: Morrow, 1992.

DON'T SIGN THE NEW EMPLOYMENT CONTRACT!

LARRY L. HANSEN

Originally Published as August 2000 feature in' Occupational Hazards' magazine.

The concept of the New Employment Contract setting forth contingent employment and hire/ fire-at-will clauses in lieu of reciprocal trust, loyalty and mutual dedication agreements is being hailed by some as a new millennium business success strategy. This thinking unfortunately falls seriously short in long-term vision, seasoned rationale, and balanced judgment. How easy it would be if American business could succeed 'globally" by managing to a simplistic "cut and slash, my way or the highway" mentality. Tom Culley, author of Beating the Odds in Small Business, says it best: "Any idiot can cut costs!" As difficult as it may be for some to accept, Dilbert isn't just a cartoon! People are real, people are a critical variable … and people do make the competitive difference!

> *"The number one workforce issue for the 21st century:*
>
> *Retaining your best employees."*
>
> - Right Management Consultants

Organizations committed to excellence need to take a longer view, one premised on the growing demographic reality that within the next 10 years, most likely sooner, knowledge-based technology and cyber-information systems will be the great levelers of international commerce ... people, will be the key differentiator of business success. Ultimately, business survival … or thrival will be linked directly to an organization's ability to optimize its human (mental, emotional and physical) resources. Business results will clearly be shaped by an organization's ability to achieve level three performance: i.e., an ability to function beyond Level 1 - 'Can-do' (ability), through Level 2 - 'Will-do' (capability), and ultimately attain Level 3 - 'Want-to' (desire). This third performance level, employees dedicated, motivated and reinforced to peak performance will be the real market force to be reckoned with in the future, both here and abroad. Timothy Butler and James Waldroop in their recent Harvard Business Review article "Job Sculpting: The Art of Retaining Your Best People" frame this reality best: "In the knowledge economy, a company's most important asset is the energy and loyalty of its people – the intellectual capital that unlike machines and factories, can quit and go to work for your competition."

"The number one lie in business:

People are our most important asset."

-LaFontaine

The New Employment Contract, inherently de-motivates employee spirit and will over the long term, and sub-optimizes organizational performance, hence becoming a self-destructing business strategy. History is clear on the issue of people and oppression … most great empires (and corporations) have fallen without a single shot being fired…they fail from within! They fail from the resistance and rebellion of their people to oppressive policy or practice.

For those contemplating adoption of these provisions into their new millennium strategies, here are some 'contractual realities' to consider prior to signing on and making these legally binding in their organizations:

"THE CONTRACTUAL REALITIES"
A/K/A
HANSEN'S LAW OF SUBORDINATE SUPERPOWER:

"For every malicious management action, there will be an employee reaction which will always be opposite … but will never be equal."

~ L. Hansen

Reality 1. Contracts entered into under duress are null and void — In many workplaces, the New Employment Contract has been imposed unilaterally rather than adopted by consensus. Such documents lack the free willed 'signature of support' of workers. Where employee buy-in and ownership is lacking … so to will be the discretionary effort critical to success. As a consequence a 'work-to-rule' performance standard will evolve and mediocrity will prevail. Excellence is rarely achieved by enforcing rules! Standards imposed will be standards opposed! People don't resist change, they resist being changed and the ultimate measure of resistance is mediocre performance.

Reality 2. Power is not the sole domain of management — As conceived, the contract fails to recognize the very real difference between 'authority' … a function of position and 'power' … a function of influence. Contrary to common belief, employees command significant power in most organizations, specifically the powers of knowledge, experience, influence, and the ultimate power; the power to "really screw things up!" The New Employment Contract does not effectively harness this power of the masses to the benefit of the organization … reference the law of subordinate superpower cited above!

205

Reality 3. "Win-win' always wins ... regardless of the odds — The provisions of the New Employment Contract run counter to emerging demographic trends which predict a severe shortage of experienced workers in the United States, starting in the year 2000 (ok, today). With a diminishing pool of knowledge and expertise on the horizon, more rather than less, competitive advantage will accrue to those organizations capable of attracting, developing and retraining experienced and loyal employees. The real winners in tomorrow's (ok, today's) hyper-competitive marketplace will be the 'we' companies ... not the 'them vs. us' organizations. Success comes from beating the external competitor ... not waging internal battles.

Reality 4. Some employees are more equal than others — The New Employment Contract is founded upon a misguided belief that all workers are equal and generic; i.e., interchangeable cogs which can be plugged into or disengaged from a process easily and quickly. The contract fails to recognize that some employees are just a bit more equal than others. These are known as top performers, and they will flee to better (more humanistic) contractual deals with organizations (competitors) who still believe in and value personnel contribution.

Reality 5. Trust counts ... a lot — The New Employment Contract erodes openness and devalues long-term contribution, hence it conflicts with the time-proven principle of "the best people make the best organization." When trust is removed from work relationships within an organization, so to is removed discretionary effort. As a consequence, so too is removed the organization's competitive advantage. The facts uncovered by Collins and Porras in their research and published findings about Visionary Companies' high emphasis on people values are just too powerful to relegate to a contractual issue.

Reality 6. Excellence is a competitive choice — Perhaps the most significant 'down-side' of the New Employment Contract is this one: "not all organizations believe it!" The contractual provisions an organization chooses to embrace as its employment policy and practices are 'issues of choice.' Those who choose the (oh, so easy) philosophy of employees as disposable assets, ultimately have to face the competitive forces of those who choose to build long-term trust relationships ... some would say no contest! I say, I'll give you odds! I'm backing my bet with the seasoned advice of some of the best experts in the field — this advice:

- W. Edwards Deming's 'eighth' point of management: Drive Out Fear so that everyone may work effectively for the company.

- Tom Peters' 27th precept for proactive managers: "Provide an employment guarantee;" and

- Michael Porter's core strategy for competitive companies: "Treat employees as permanent instead of employing demoralizing hire and fire approaches."

The literature on management, human relations and excellence is clear; a strategy that sub-optimizes human contribution to the business process is a strategy doomed to long-term failure. The unilateral devaluing of 'people' and their critical assets (experience, expertise, loyalty, knowledge and trust) as embodied in the context of the New Employment Contract leads to this conclusion.

And … what does all this emphasis on human relations and positive management practices have to do with safety? — Just about everything!

Hank Sarkis, MBA, President of the Reliability Group, an Organizational Behavior consulting firm has been collecting data which reflects a strong relationship between these factors. His database of organizational factors and accident experience has been evolving over the past 15 years and the emerging correlations are insightful: "Over the past six years, we have identified approximately 80 workplace variables that have significant statistical relationship to accidents. The most significant to date are:

#1 - Workplace stress … and

#2 - Cheerfulness in the workplace (Sarkis)

Dr. Deming was indeed insightful when he acclaimed "Export everything…except American Business practice!"

REFERENCES

Aguayo Rafael, Dr. Deming - The American Who Taught The Japanese About Quality, Simon & Schuster, New York, 1990.

Bernander, Terry, Empowering Workers Fuels A Corporation's Turnaround, *Occupational Health and Safety,* June 1996.

Butler, Timothy and Waldroop, James, Job Sculpting: The Art of Retaining Your Best People, *Harvard Business Review*, Sept. – Oct. 1999.

Tom Culley, Smart Owners Know Basics Still Matter, *Knight Rider News Service*, Syracuse Herald.

Hansen, Larry L. and Coderre, Paul, Twelve (12) Unlegal Ways to Slash Workers' compensaiton Costs, *Professional Safety,* June 1997.

LaFontaine, Paul, The Five Most Common Lies in Business, *Fast Company Magazine*, Aug./Sept. 1997.

Tom Peters, Thriving on Chaos, Alfred A. Knopf, New York, 1987.

Michael E. Porter, Competition in Global Industries, *Harvard Business School Press*, Cambridge, Massachusetts, 1986.

Right Management Consultants, Leaders Identify 5 Workforce for the 21st Century, *Leadership Strategies Newsletter*, June 1997.

Sarkis, Hank, Near Miss: *20/20 Foresight, Employee Assistance Newsletter,* July 1991.

Simon, Rosa Antonia, The Trust Factor in Safety Performance, *Professional Safety,* October 1996.

SAFETY EXCELLENCE REQUIRES LEVERAGING THE BEST OF PEOPLE

LARRY L. HANSEN

An essay on the critical role of human relationships in achieving safety excellence.

When business managers set forth to improve operating results, they traditionally focus on the R-words; i.e., Reorganization, Restructuring, Rightsizing, Reengineering and RIF'ing (reduction in force) of staff. These efforts target two key elements of the business process; organizational structure (arrangement of the boxes) and technology (automation of the process). These factors drive efficiency; i.e., making things leaner, cleaner, faster, flatter and wider to improve the organization's ability to do things right.

Although these box reconfiguring, put terminals on desks approaches to improving performance have reduced costs, they have fallen short in their goal of delivering greater margin. Such initiatives often lead only to an organization becoming more efficient at inherently flawed processes. Although operating expenses are reduced, primarily through staff cuts, other operational costs driven by systemic reactions arise to impact productivity and negate margin contribution.

A focus on efficiency (changing structure and process) alone is not a complete strategy. No matter how the boxes are flattened, squeezed, scrunched, or reduced in number, the fact remains, little changes; all we get is more or less of the same! The truth of the matter is a focus on the R-words alone, does not appreciably improve the Y-words; i.e., the results generated from the business process: quantity, quality, reliability, productivity, creativity, delivery, customers happy, safety and yes, ultimately… profitability!

If an organization has performance problems with any of these Y-words (outcomes), the causes are imbedded in the business process and management practices of the organization. Therein lie the common causes of both business success and failure, no matter what measure is used.

James Champy, co-author with Michael Hammer of "Reengineering the Corporation" now, some 10 years later, in his new book, "Reengineering Management" retrospectively addresses the shortcomings of the earlier reengineering revolution. He concludes: "Although the jury is still out on 71 percent of the ongoing North American reengineering efforts, overall the studies show participants fail to attain their benchmarks by as much as 30 percent. If I've learned anything, it is that the revolution we started has gone, at best, only half way." (Champy)

What's missing? Why have improvements not reached the bottom line? The answer, missed ten years ago, is clear today, -- People … and Values!

The reengineering movement focused primarily on structure and technology; it omitted the critical human element of the business process. Rather than reconfigure the boxes we needed then, and continue to need now, to change what's inside the boxes; i.e., the culture and values of the organization. These are the human element issues which most impact the business process, influence performance, and determine the ultimate outcomes, including safety outcomes, of an organization. Champy identifies four critical issues, which must be addressed for reengineering of management to truly succeed. They are:

• **Issues of Purpose** – What is it that we're in business for? (The purpose beyond profit.)

• **Issues of Culture** – How do we overcome fear in the organization and generate an environment of willingness and mutual confidence?

• **Issues of Process and Performance** – How do we get the processes we want and the performance we need from people? … and

• **Issues of People** – Who do we want to work with; how do we attain them and how do we keep them? (Champy)

In a commentary on the competitive state of manufacturing in the United States, Jack Welch, Chairman and CEO of General Electric Corporation, confirmed the opportunities for leveraging the human side of business enterprise: "We've spent the majority of our energy in the '80s working, appropriately, on the hardware of American business, because the hardware had to be fixed. The Japanese, on the other hand, have the software, the culture that ties productivity to the human spirit--which has practically no limits. That's where we have to turn in the '90s, to the software of our companies; to the culture that drives them." (Oakley and Krug)

Research into organizational performance and operational excellence such as that done by Tom Peters in Search for Excellence, and Collins and Porras' in Built to Last, consistently find a strong relationship between an organization's culture (it's basic beliefs and values exhibited in its leadership practices) and the results (Y-words) that it produces. Business success, be it measured by productivity, revenue, market share, safety, or ultimate profitability, is the outcome of a strategically designed seven step process which pivots around the human element of the organization.

THE OPERATIONAL EXCELLENCE PROCESS

Operational excellence, i.e. excellence in all measures of business performance, is the outcome of an organization committed to:

- **Best Principles** – A commitment to excellence values — Setting a high bar.

    A high value placed on the health, safety, and well-being of workers.

- **Best Processes** – A dedication to operational 'effectiveness' — Doing right things.

    A commitment to proactively designing safety into key processes and operations.

- **Best Practices** – An emphasis on operational efficiency and continuous improvement. - Doing things right.

    A high importance placed on defining, observing and reinforcing safe behavior.

- **Best of People** – Leadership practices that unleash discretionary behavior.

    A recognition that how managers manage, directly impacts how workers perform.

- **Best Performance** – Systems that maximize human, mechanical and environmental interactions.

    A commitment to planning and developing systems that assure safe outcomes.

- **Best Productivity** – A scorecard that balances 'hard and soft' measures, and values soft more.

    Inclusion of safety metrics (activities and results) in the business measurement system.

- **Best Profitability** – Financial outcomes over time superior to the industry.

    An ability to demonstrate safety's value and contribution to shareholder value.

Note that this 'excellence process' (safety or otherwise) does not specify 'the best people' but rather is leveraged by the concept of the 'best of people.' I have yet to see the following contention disputed: All employees are created equal, they just don't perform that way. And, we must ask: Why?

Productivity surveys done by Organizational Performance consultant Daniel Yankelovich confirm: "About a fifth of workers feel they're giving everything they can to the job; another fifth don't want to give anything more, and the remaining 60% percent say they would give more if there was more in it for them" (Oakley and Krug). Human Resource specialists have worked diligently to build effective screening procedures to cull out low performers and assure a flow of high quality workers into our organizations. Assuming these processes are reasonably effective, and in view of the fact that safety expectations continue to be less than excellent, odds are the problem(s) lie in our values and leadership practices, more than our people.

The DuPont Corporation provides an excellent example of safety values and leadership. A few years back, at an ASSE western New York PDC I was asked to host a luncheon discussion with participants which included a manager from a local DuPont facility. Always having had a special interest in the mystique and exceptional results of DuPont, I asked about that plant's safety performance. Midway through a description of a long running "no lost time' incident record, one of the participants asked where that plant got its people, firmly believing that the plant must have had some secret staffing source different from all others in the area. He couldn't believe that any organization could achieve such results using the same labor force as others in the area.

In reply, the DuPont representative politely, yet pointedly, dismissed that notion with this response: "We get our people from the same place you do … the only difference when it comes to safety, is: **we mean it!**" At DuPont, safety truly is a core value and the actions of all managers and employees bring it to life. DuPont associates understand, accept and embrace the fact that safety is what matters most, in their process, in their methods, in their products, in their performance, and ultimately in their operational results.

The simple truth is that leadership practices either suppress or unleash the best of people in safety and all other aspects of operational performance.

One of the most frequently asked questions in our profession is: How does an organization become world-class, i.e. "first in its class in safety performance?" That answer is quite simple. Excellence companies…put safety first! It's DuPont's answer, and also that of another industry leader, AK Steel. It's interesting to see how this organization presents itself to the public and how they emphasize 'first things first' when you visit the operations section of their web site. I encourage you to visit them (www.aksteel.com); it's a journey worth taking (three guesses what comes first!).

Bottom line: The common denominator of excellence companies is leadership; more specifically, the leadership practices (not administration, manipulation, or management), but rather the leader actions which bring out the **best of people.**

'BEST OF PEOPLE' STRATEGIES AND PRACTICES

Here are 12 **'best of people'** strategies for optimizing human performance and achieving world-class safety performance. Do these things routinely, repetitively and sincerely, and operational results will improve…how could they not?

1. **Hire by values … and let those who you value do the hiring**. Involve employees in the selection process and set a high bar for all candidates. How does an organization get strong values? … It hires them!

2. **Invest generously in employee development.** Continuous learning and renewed skill building is critical to ongoing success. Recent findings from the National Research Council cite the half-life of worker skills now to be three to five years! We need to replenish these at a rate faster than utilization. (Prichett)

3. **Foster teamwork, not teams.** Encourage cooperation by sharing challenges, sharing information and sharing the rewards that produce success. The key challenge is to make the 'them' in them vs. us … the competition!

4. **Build a change friendly, continuous improvement environment.** Place high value on individual and group innovation, creativity and reasonable risk taking. And, be willing to reward well-intended efforts that falter — it's the effort that counts. You will get more of what you reward.

5. **Communicate openly and aggressively.** There's no such thing as a 'communications gap' in business. If managers fail to fill the communication pipeline with accurate and positive information, subversive information will seep in to fill the void in the form of 'a grapevine'. You can never communicate enough — double your efforts immediately!…(then multiply that by two).

6. **Employ positive reinforcement over progressive discipline.** A punitive work environment does not increase performance; only reinforcement permanently elevates the level of desired behavior in an organization. The behavioral sciences confirm a 'positive to punitive' feedback ratio of 4 to 1 to be necessary for improved work performance. In safety, the key is 'P.O.U.R.' it on, provide: 'Positive Only Unconditional Recognition'!

7. **Reward right efforts … even if they fail.** Recognize and reward right thinking and right risk taking for this challenges the status quo, encourages forward motion, and ultimately leads to breakthrough successes. Remember even in the pros, a great batting average is only 300! — A 70% failure rate!

8. **Celebrate earned successes.** Share the spoils of success with those who contribute to its achievement. The purpose of celebration is to relive the journey and frame those events and individuals that produced

the success. The key is to drive these memories deep into the values of the organization — we remember the fun things most.

9. **Listen, listen, and listen some more — did you hear what I said?** People throughout an organization have the answers. People are not the problem; people are the solution ... it's critical that we hear their message...and no, I'm not recommending a 'suggestion box!'

10. **Believe in people.** When you believe in people, and show confidence based on those beliefs, most will rise to the challenge. By raising the bar, you elevate the performance of most and the results of all ... "you can do it, I know you can!"

11. **Pay attention to what matters most.** Paying attention to something increases its output. Elton Mayo and his research team demonstrated this principle at the Hawthorne Western Electric Facility decades ago, and it's equally as true today. To get greater results, pay greater attention to the people who produce them... and, above all else,

12. **Tell the truth!** Truth is the mortar which strengthens trust and bonds employees to the organization. The sequence is simple: from truth flows trust, from trust flows credibility, from credibility flows performance, and from performance, flow results.

These 12 principles, and the seven-step operational excellence process identified earlier, pivot around a critical center point... the '**best of people.**' I don't believe it was pure coincidence that Dr. Deming's central (7th of 14) point of quality management was: "**Adopt and institute leadership aimed at helping people do a better job.**" (Deming) People are the core of an operational excellence process and the key to excellence results! Can we possibly believe that safety excellence is the outcome of any thing different... regulations and safety programs perhaps?

REFERENCES

James Champy; Reengineering Management, The Mandate for New Leadership (New York: *Harper Business Publications,* 1995)

W. E. Deming; Quality, Productivity and Competitive Position (Cambridge, MA: *MIT Center for Advanced Engineering Study,* 1982)

W. E. Deming; Out of the Crisis (Cambridge, MA: *MIT Center for Advanced Engineering Study,* 1986; Cambridge, MA: Cambridge University Press, 1988)

Carol Kennedy; Instant Management – The Ideas From The People Who Have Made A Difference in How We Manage (New York: William Morrow & Company, Inc., 1991)

Doug Krug and Ed Oakley; Enlightened Leadership Getting to the Heart of Change (New York: Simon & Schuster Publishing, 1991)

Price Pritchett; Carpe Mañana, Tomorrow's Coming, Ready or Not; Pritchett Rummler-Brache, Plano, TX

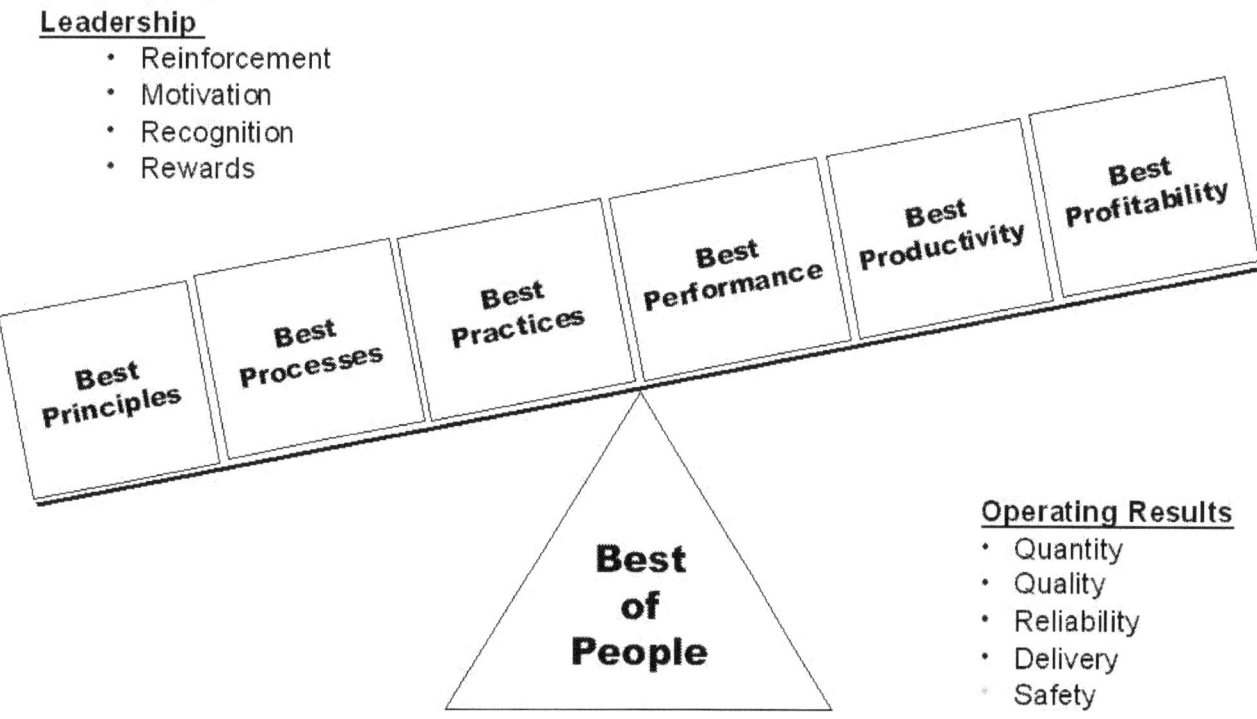

A UNIVERSAL MODEL FOR SAFETY X-CELLENCE.

SECTION 10

Communications

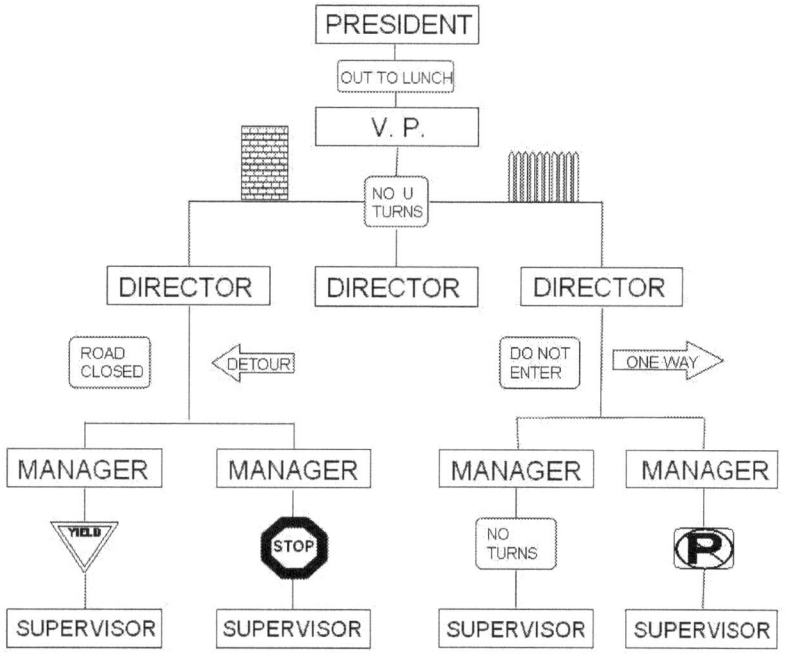

"Organizations cannot operate intelligently if those with power are cut off from those with knowledge."

-David Francis

The Universal Model of Safety Excellence

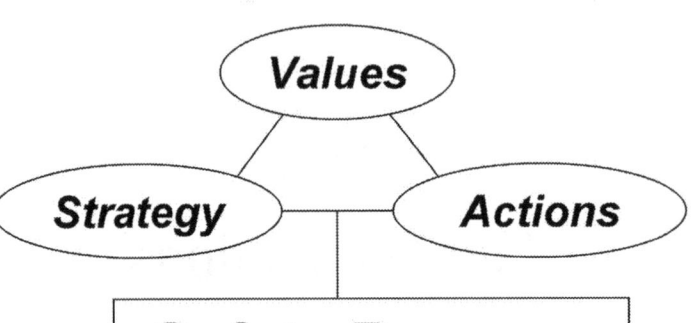

Safety Leadership		

Values

Strategy   Actions

Safety Process

Integration

Safety Management		
Human Relations	Communication	

A QUESTION OF EXCELLENCE:

COMMUNICATIONS
Have we created information systems that facilitate safe decision-making in our organization?

EFFECTIVE SAFETY COMMUNICATIONS
PART I

WHY WON'T THEY LISTEN?

LARRY L. HANSEN

Originally Published as September 2004 feature in' Occupational Hazards' magazine.

In Part I of his series on safety communications, loss control expert Larry Hansen reviews why so many safety communication efforts fail, and shares three truths about effective communications.

Ask any group of business managers to name the one organizational problem that frustrates them most, and which is often at the core of operational snafus, and I'll bet more often than not you'll hear: "POOR COMMUNICATIONS!" And that goes for accidents, injuries and LOSS costs as well. The National Council of Compensation Insurers has estimated that as much as 40% of workers' compensation costs can be linked to how an employer communicates with, and responds to worker injuries.

Communications expert John Drebinger says: "You attain the next level of excellence by changing who you are and you change who you are by changing what you do." When it comes to designing, implementing, and administrating an effective safety communications process, researchers suggest that many organizations need take a closer look at 'what they do!'

With few exceptions, poor communications stands out in both performance literature and management practice as the #1 cause of substandard performance in many organizations. It's cited frequently for: low productivity, late delivery, poor quality, low moral, workplace accidents, and most other undesirable events impacting profitability. In "A Passion for Excellence", Tom Peters and Nancy Austin concluded that: "Nonfunctional communication is at the heart of most management problems." And most operations executives agree, allocating millions annually to improve the situation. Yet in spite of this recognition, and huge financial commitment, communication problems persist.

In titling this article, I actually faced this exact dilemma. I'll bet most readers, to this point, assumed I was referring to 'employees' as those 'who just won't listen'! Oh well, another case of poor communications! The title, and theme of this article, addresses effective communication as a management opportunity, not an employee problem. It's managers that need to listen (and learn) say researchers and management experts, including Stephen Covey, author of "The Seven Habits of Highly Effective People", who identifies communication as the fifth habit: "Seek first to understand, then to be understood."

When it comes to communication and credibility, employees in America's workplaces have some very definite opinions, and they're not very complimentary…and not too far from wrong. Spend a fair amount of time in employee lunchrooms, break areas, or by the office water cooler, and eventually you'll hear this all to common satirical question: "How can you tell when an executive is lying?" Answer in unison: "Their lips are moving!" Would I lie to you? According to the findings of an ASCLU workplace survey published in the August 1997 issue of Fast Company magazine, the answer is --'Yup!' Of those polled: 93% acknowledged that they lie regularly; 56% admitted they have been pressured to act unethically, and 48% have committed illegal acts. And a more recent pulse on the issue— Dec. 30, 2003; NPR's website listing the top ten 'strategy and business' books is introduced this way: "Not surprisingly, books about scandals dominated the best-selling business books of 2003. Perhaps the Budweiser Lizard was insightfully correct in his observation: "Never send a weasel to do a ferret's job!" (Oh, how I love that commercial!) Business leaders indeed have a credibility gap in the eyes of many frontline workers, and this perception (reality) is more frequently influenced by the communication techniques they employ, rather than the content of the message they attempt to convey.

Research on organizational communication suggests that the majority of advice given to business executives on how to effectively communicate with frontline employees is 'wrong!" To move information through an organization, many companies establish corporate communications departments which focus on 'word-smithing, format, and technology' to facilitate internal messaging. The answer to many organizational communication problems, however, doesn't lie in the message content or the technology, but rather in the methods employed.

Think about the traditional communication mindset in your own organization. What methods are most frequently used by upper management to communicate safety information to frontline employees? Make a list of these techniques, and then arrange them in order of perceived effectiveness. This exercise might seem simplistic, but it, in fact reveals the prevalence of what I call the: "I've got the bull by the horns" phenomena -- the belief that senior management understands the information needs, and controls the messaging process.

At this point, let's 'cut to the heart' of the issue. If we want to find out what is most important to, and how to most effectively communicate with frontline employees, whom should we ask? Who would know best? Who are the experts?

EMPLOYEES, of course! In doing this, we discover a very different perspective, the "I've got the boss by the hands" phenomena a/k/a the truth! Bottom line, if we're not communicating successfully with employees on safety issues, it's most likely a matter of technique and approach. We need to determine why our messaging fails, and change our approaches accordingly.

As Einstein (paraphrased) once said: " The thinking that gets you out of trouble must be far different than that which got you into it in the first place" -- although you wouldn't think it would take a genius to figure that one out!

When it comes to effective communication, what we desire, and what we in fact attain are indeed very different. Ask a bank teller, computer programmer, machine operator, sales clerk, construction worker, truck driver, or office worker and you'll quickly learn that frontline employees of all titles and roles:

1. Distrust information they receive from senior managers;

2. Believe that company publications are propaganda;

3. Hate watching executives on videos;

4. Don't bother reading those annoying leaflets in their pay envelopes, and

5. Have little interest in receiving general corporate information.

Contrary to common beliefs, frontline employees have little interest in receiving anything that doesn't directly impact them or their work. If a company's communication program isn't targeted directly to the worker and his or her specific interests, it's wasted paper (or e-time). Most isn't . . . and hence most is!

The critical question then is, "what works?" Much research has been conducted to determine that, and the findings are enlightening. A common corporate assumption is that employees want information directly from senior managers via mass techniques such as satellite telecasts, corporate meetings, road shows, executive stump speeches, and feature articles in the company newspaper - NOT!

These approaches are founded on the 'as-the-bull-sees-it' approach, which assumes that employees perceive senior executives as well informed and credible. An unfortunate reality, however, is that employees frequently distrust senior managers. Employees trust one person in their work environment most -- their supervisors; the person they work with daily and communicate with routinely. Employees view their immediate supervisor as their most credible source of honest and objective information. In view of this, they prefer to receive, and most readily accept communication from them. Why? Because they trust them.

Evidence supporting this includes: (Larkin)

• A Wyatt Consulting study (England) finding that 70 percent of employees felt the information they received from management was misleading; and,

• A Lou Harris & Associates Poll (United States and Canada) revealing that 62 percent of office workers were not confident that management was honest in their dealings with them.

And, in contrast:

221

- A major US auto manufacturer's study confirming that 83 percent of employees rank their immediate supervisor as their most believable source; and

- Australian researcher Dennis Taylor's findings that 90 percent of Australian employees believed their supervisor was either 'always or normally' telling the truth.

From these studies by researchers Sandar and T.J. Larkin, three 'truths' have emerged as guide posts for designing effective organizational communications for safety:

Truth 1 – The most effective communication to frontline employees comes directly from their immediate supervisors.

Supervisors are the opinion leaders for frontline employees. They are the ones that have earned credibility, and hence, are the ones that can most influence employee behaviors. A top or mid-level manager's success in communicating on safety depends upon their success in communicating with supervisors!

Many corporate communication initiatives are inconsistent with this fact. In practice, communication is dominated by 'top to bottom' approaches; i. e., (open-door policies, MBWA, direct lines to the boss, etc.) which creates supervisory bypasses, weaken supervisory credibility, and result in a 'not so good news/bad news' scenario.

The 'not so good news' is: employees take this stuff seriously! When told they have a 'direct line' they use it, and problems, concerns, and questions 'rise up' and inundate executive offices, resulting in organizational overload and slow response, which further damages management's credibility and fuels employee cynicism.

The 'bad news' is: when executives do respond to employee concerns on a timely basis and fix problems, the message sent to frontline employees is clear: "Hey, if you want something done around here, you gotta take it up top!" (The ultimate bypass: front line to top line.) It's the start of clogged communication arteries and organizational paralysis!

CEOs embrace communication approaches such as these because they truly believe they favorably impact employee behaviors. An A. Foster Higgins study of 164 Fortune 500 CEO's found that 97 percent felt their communications affected employee satisfaction, and 75 percent believed it impacted job performance. Evidence, however, confirming that direct communications from executives to frontline employees positively impacts work behavior is slim, while there is a wealth of evidence to the contrary.

To impact safe employee behavior, managers must increase the perceived power of supervisors by communicating credible information to them (i.e., putting them 'in the know') and by establishing them the 'credible source' for employees. Bottom line, it is not an employee's relationship with a CEO that impacts safe work performance; it is the employee's communication relationship with their supervisor that's most important. A strong credible relationship at this level minimizes distortion.

Studies confirm that somewhere between the CEO (sender) and frontline employees (receivers), organizational messages are distorted, and the 'snag' is generally middle management for a myriad of reasons. Because of this, executive messages don't 'trickle down' as conceived, nor do they have the impact on performance expected. This supports Larkin's second communication truth:

Truth 2: Effective communication must be delivered 'face-to-face' by supervisors.

Dick Vitale would say: "Get in their face, baby!" . . . and he'd be right because supervisors are the information sources most recognized and trusted by employees.

Unfortunately, executives often don't see this and continue traditional communication campaigns, publications, video reports, and meetings. The problem is what executives have to say generally doesn't interest frontline employees. I've seen this first hand, perhaps you have as well? One company with which I am familiar tried to communicate corporate challenges and results to regional staff by sending the President and a contingent of senior staff on a 'dog and pony show' every spring to deliver a 'state of the company' message to all regional employees. These sessions always ended with: "Are there any questions?" There never were, and not because the meetings were highly insightful or informative Zzzzzzzz! To avoid embarrassment local managers would 'plant' questions in the audience. Therefore, they continued.

Researcher Larkin identifies British Telecom an interesting case in point. British Telecom is reported to have one of the most respected corporate communication programs in the world; consisting of numerous proactive initiatives, yet their results in 'informing their work force' were only slightly better than the United Kingdom's average (i.e., British Telecom 42 percent of employees well informed versus the United Kingdom's average of 40 percent.). Why did all of this effort produce but a small difference? Larkin suggests it was because it conflicted with communication truths #1 and #2 – it wasn't delivered by supervisors and wasn't done 'face-to-face.'

Truth 3: Frontline employees are only interested in information that directly affects them, their job, and their immediate work.

This is more commonly referred to as: 'WIIFM' (What's In It For Me). If management is not broadcasting to these 'call letters', employees aren't going to tune into the message! Executives can talk about risk patterns, loss trends, accident rates, insurance costs and corporate strategy, but to the frontline employees, that's all irrelevant! Employees don't think like managers because employees aren't managers! Employees have different beliefs, concerns, and interests. Employees don't have bad attitudes -- they just don't value the same things managers do. CEOs need to understand that 'how they communicate', more than 'what they communicate', is the critical issue at least to those with whom they are trying to communicate.

Employees interpret information that isn't specifically targeted to them, about them, or for them with cynicism. As far as they're concerned, it's 'BOHICA (Bend Over, Here It Comes Again!')

Not typical? An exaggeration? Not in this organization you say? Unfortunately, it's truer than most would like to admit! Research confirms the reality of distrust and cynicism between management and employees: (Larkin)

*   A Hay Management Consulting Group (Canada) study:

    "Only 50 percent of employees have pride in their companies";

*   A Hay Group study in the United States:

    "70 percent of hourly employees lack confidence in top management";

*   A Boston University School of Management survey:

    "80 percent of workers felt management could not be trusted."

Consider the reality (i.e., employee perceptions) in your own organization? Have your company's best efforts to communicate a sincere interest in safety improvement ended up with cynical employee responses, such as: "SAFETY IS OUR #1 PRIORITY" changed on the work floor to read: "SAFETY IS OUR #1 SLOGAN?"

Executive intentions, although admirable, aren't good enough. What matters in effective communications is an understanding of what information employees need and want. Employees want to know how they're doing, what changes are planned, and what it means 'specifically' for them. Employees need feedback that is relevant to their work environment and which will help them perform better.

If managers want to improve safe behavior, they must shed a number of traditional communication beliefs and approaches, and invest in building a communication process guided by the three core 'truths' of effective communication:

1.   Communicate through supervisors (to build strong first line trust and credibility),

2. Require that communications be 'face-to-face' (to allow for feedback, expression and dialogue), and

3. Communicate information that is specifically relevant to employees and their work (address their WIIFM questions).

Watch for our October Occupational Hazards issue where our Part II in our journey to effective communications continues…to the strange new worlds of 'NETMA' and 'TMSIADK'

REFERENCES

Albrecht, Karl, The Only Thing That Matters, Harper Collins, New York, 1985

Austin, Nancy and Tom Peters, A Passion for Excellence, Random House, New York, 1985

Barrack, Martin K., How We Communicate - The Most Vital Skill, Glenbridge Publishing Ltd., 1988

Covey, Stephen R., The Seven Habits of Highly Effective People, Simon & Schuster, New York, 1989

Mansdorf, Zack, Communication - The Key to Success, *Occupational Hazards Magazine*, May 1993.

Larkin, Sandar and T.J., Communicating Change - How to Win Employee Support for New Business Directions, McGraw Hill, New York, 1994

Looking Up - Survey of Workers' Compensation Issues, *Risk and Insurance Magazine*, November 1994, pp. 30

O'Reilly, Brian, The New Deal - What Companies and Employees Owe One Another, *Fortune*, June 13, 1994, pp. 44-52

Werther, Dr. William B., Dear Boss, What Every Manager Needs to Hear and Every Employee Wants to Say, Meadowbrook Press, New York, 1989.

EFFECTIVE SAFETY COMMUNICATIONS
PART II

HOW WILL THEY KNOW?

LARRY L. HANSEN

Originally Published as October 2004 feature in' Occupational Hazards' magazine.

In Part II of his series exploring the universe of effective safety communications, Larry Hansen examines the seven elements of organizational communication and their critical role in shaping safety success.

W. Edwards Deming believed that an organization's ability to achieve 'Quality Excellence' hinged upon management's ability to answer one simple question; the same question that lies at the core of 'Safety Excellence", it is: 'How will they know?"

How will they know…what is expected? How will they know…what process to use? How will they know…if the process is working? And most importantly, how will they know… when excellence has been achieved? In this short yet insightful question, Dr. Deming captured the critical importance that 'data, information, and profound knowledge'; a/k/a 'effective communication' plays in attaining operational excellence. Deming further noted that: "Numbers are numbers; numbers are not knowledge", cautioning managers to seek 'the good reasons for poor performance'; i.e., the reasons behind the numbers. Attaining 'organizational intelligence' requires that an organization develop systemic ways to obtain objective data, identify process variances, and solicit unfiltered feedback on cause and correction, three elements critical to an effective communication process.

Many businesses unfortunately have a bias for 'quick and easy', not necessarily effective communication processes. They focus on moving information, not conveying understanding. In addition, today's 'e-com' world has fueled information overload that further threatens organizational intelligence. Consequently, employees in poor communicating organizations feel that they live on planet 'NETMA' – (Nobody Ever Tells Me Anything!). Ignorance may be bliss…but in business and risk management, it's what you don't know that can hurt (and will cost) you!

A reader survey conducted by Human Resource Executive and Risk and Insurance magazines found that although business executives recognize 'employee communications' to be a low cost way to address workers' compensation concerns, only 39% of those organizations surveyed had an on going communications program. Perhaps this is the reason that Tom Lynch, past executive of Lynch Ryan Associates contended: "Show me a company with high costs, and I'll show you a company where employer and employee do not communicate effectively."

And, what's the net effect of poor communications on employee attitudes and work performance? Cynicism, suspicion, and credibility gaps between executives and front-line workers wherein trust diminishes and performance declines. To succeed in safety, process leaders must improve the level of 'corporate intelligence' in their organizations through communication systems that work hard at working!

To optimize safety performance, an organization must examine the ways safety is addressed in key communication systems and integrated into the organization's vision, values, and mission. If safety is to evolve beyond programs to become part of the organization's culture; i.e., 'how business is done', it must be included in and reinforced by all key communication systems. It must be embedded in the organization's:

- **Core Guidance Documents** – Its written (and electronic) directives, specifically its Vision, Purpose, Mission, and Values statements.

- **'One on One' Contacts** – Its routine manager contacts and interactions with individuals and direct reports.

- **Formal Meetings and Training** – Its formal gatherings and scheduled skill building and information exchanges.

- **Feedback Systems** – Its systemic process of listening, hearing, gathering, assessing, and acting on critical performance information.

- **Measurement and Metrics** – Its reports, scorecards, and performance spreadsheets.

- **Recognition and Rewards** - Its 'WIIFM' (What's In It For Me?) performance management systems, and

- **Management Actions** – Its executive decisions, and management practices.

If any of these means are omitted, or compromise safety as a value, employee cynicism grows and performance declines. Let's examine these seven elements of organizational communication and their critical role in shaping safety success:

1. PRINTED (PAPER OR ELECTRONIC) GUIDANCE DOCUMENTS

If employee safety is to be important in an organization, it must be reflected in its written guidance documents and corporate position statements that define the corporation's basic beliefs and values. If senior management doesn't put safety on the road map to success, the organization will never travel down that road.

227

Two of the most important documents are the corporate Vision and Mission statements.

What is 'notably missing' from this statement?

MISSION STATEMENT

"The mission of 'ABC/XYZ' Corporation is to meet and exceed the goals of our clients.

We are dedicated to delivering the highest quality of service for every project and every client. We are a customer-oriented company, and strive to create positive relationships with owners, sub-contractors and suppliers. We will continue to utilize the latest technologies and innovative techniques to ensure the satisfaction of our clients."

And, what is 'most noticeable' in this 'Vision Statement of PSI Energy?

PSI VISION

"WE WILL BE A LEADER IN THE EMERGING ENERGY SERVICES INDUSTRY BY CHALLENGING CONVENTIONAL WISDOM AND CREATING SUPERIOR VALUE IN A SAFE AND ENVIRONMENTALLY RESPONSIBLE MANNER."

And, what is unmistakably noticeable in the unique way Scot Forge, an industry leader in Spring Grove, Illinois has visually integrated safety into their corporate Vision & Mission:

SCOT FORGE VISION & MISSION

S – Safety

C – Continuous Improvement

O – On time delivery

T – Total customer satisfaction

FORGE
At Scot Forge, Safety comes first! And success follows.

POLICIES, RULES AND SOPS. Many organizations devote considerable time, and effort to the development and administration of rules, while they invest little in building trust. As a result, workplaces are policed by 'rules' rather than led by 'values'.

These organizations manage by the '95 – 5' rule. Ninety five percent of the rules they promulgate (including safety rules) are done in response to the inappropriate actions of but five percent or less of the work force. Consequently, rules are ignored by the 5 percent for which they were intended, and alienate the other 95 percent who never needed them.

We need to start building values for safe performance rather than sending out addendums to rulebooks that are rarely read and generally not heeded by those who violate them. In companies that are over managed and under led, safety is a game of chance; "Catch me if you can." Employees don safety equipment and follow the rules while they are being watched, but as soon as supervision or the safety manager leaves, off it comes, and back to unsafe practice they go. In organizations that invest in building trust through the creation of 'shared values', something different exists…it's called a safety culture. People behave safely because it's the right thing to do…for them…their peers… the company …and the stakeholders in the business. Behavior is guided by values and reinforcement, not bludgeoned by rules and punishment.

2. INDIVIDUAL AND GROUP CONTACTS – Studies by NIOSH, Boeing, and the Reliability Group, an organizational performance consulting firm have all identified the impact of 'employee satisfaction' on the level of safety in a workplace. The data developed by the Reliability group has in fact, determined that the number one predictor of a safe vs. unsafe workplace is employee cheerfulness and satisfaction. And a key factor in determining employee satisfaction? Supervisors!

Unfortunately, human resource studies have also found that most supervisor/employee interactions are negative with the number of disciplinary notices sent to employee personnel files far exceeding the number of positive entries. There's also a growing body of evidence linking stress in the workplace to cases of deviant employee behavior…including violence! And, a primary source of this stress and often the target of stress induced violent behavior is 'the boss'! All this suggests that 'work should be fun'…but we just don't see this performance requirement in many managerial job descriptions.

3. MEETINGS AND TRAINING – Much has been written about the 'learning organization'. What does that mean? Quite simply, it means an organization committed to the ongoing development of the organization (all employees) at all levels. William Lareau, author of the American Samurai claims that workers in Japan and many European countries are 'smarter' than U. S. workers…in some cases two to three times smarter. But, he acknowledges that this has nothing to do with individual intelligence. They are two to three times smarter, because they receive two to three times more training!

In the first year of the Clinton administration, then Labor Secretary Reich proposed the '1.5% Solution', an initiative which would have encouraged American business to invest 1.5% of payroll in employee training. It remains a solution not applied. In a number of foreign countries, it's not unusual for 3% to 5% of payroll to be dedicated to employee development. Moreover, in more advanced organizations, funding is based on 'revenue and/or profits', not just payroll. Performance literature has reported that excellence companies spend 2.5% to 3% of payroll on training We need ask: "What percentage of payroll is allocated to employee training in our organization? An unfortunate reality is…not many managers know.

4. FEEDBACK SYSTEMS (FORMAL & INFORMAL) – Feedback systems allow problems to surface early, and provide a means for management to continually strengthen and reinforce the organization's core beliefs and values.

Feedback systems in many organizations often fail…not due to design, but rather due to administration. I recall one clear example. A health care organization attempted to improve its employee involvement process by installing an employee safety suggestion system. Suggestion boxes were placed throughout the facility, and employees were encouraged to provide input, which was reviewed and published monthly with recognition…an excellent concept. The program worked extremely well in its early stages, but over time, feedback diminished, suggestions decreased, and ultimately the program was discontinued. Frustrated by the apparent employee apathy, the administration engaged a consultant to help determine why the program ran out of steam, and how to rekindle employee interest. The very first effort to discover these answers involved meetings with employees to gain their perspective. It was quickly determined that the short-lived success of the program had little to do with a lack of interest or motivation. Employees were quick to advise: "There never were any pencils!"

And speaking of opportunities, don't overlook the opportunity to seed and feed the 'informal' communication systems in your organization… a/k/a 'the grapevine'. Most organizations regulate and/or place strict limits on employee interactions, such things such as coffee breaks and interoffice visits. Other organizations see these informal employee interactions in a very different way; i.e., as a chance for employees to exchange useful ideas at an operating level. I guess it's all about 'trust' and how you look at it.

5. MEASUREMENT AND METRICS – What is counted in an organization sends a clear message to employees as to what's important. If an organization proclaims safety important, but only measures production, quality and delivery, employees quickly conclude what can be compromised at the end of the month when shipping volume is down and quotas have to be met… "To hell with the guard…we gotta ship 10,000 pieces!"

An automotive components manufacturer with an extremely high accident rate and excessive workers' compensations costs driven by an Exp. Mod. of 2.60, engaged me to assist them in solving a LOSS problem that now was now threatening their ability to compete for future projects. This organization was big on visible measurements on the production floor. At the beginning of each shift, large scorecards were posted at the beginning and end of each process line displaying production and quality quotas for that shift. These were updated hourly by shift supervisors, posting production volumes and QC metrics against goals. This organization consistently met production and quality quotas…but failed to meet productivity (production/

costs) goals...any guesses why? Indeed, "What gets measured indeed gets done."

6. RECOGNITION AND REWARDS – 'R&R' provides the strongest means for management to express commitment and communicate the values (what's really important) in an organization. What get measured gets done...but what gets measured and rewarded gets done well! There are 'two' very common reasons why at-risk behaviors exist in many organizations. Hint: the 'R-words'...failure to provide appropriate 'Recognition and Rewards' for desired behavior. In her book, "Unlock Behavior; Unleash Profits" Dr. Leslie Wilk-Braksick concludes: "The primary tool for unlocking discretionary effort in an organization is the application of positive reinforcement for desired behaviors."

A William M. Mercer, Inc. survey of 200 communications managers some years back determined that 'recognition for a job well done' was the top motivator of employee performance. It graded 4.9 on a scale of 6.0, surpassing money at (4.8) and job challenge at (4.3). The most powerful motivator of employee performance can be reduced to one word...THANKS, and costs next to nothing to deliver, yet positive reinforcement compared to the use of progressive discipline, it is a relatively 'usedless' performance improvement strategy in many organizations.

7. MANAGEMENT ACTIONS – Executive actions and manager practices are the ultimate communicators of how much an organization values the safety of its people. The three most powerful ways of establishing the importance of safety in a workforce is: by example, by example, and by example. Or as Tom Peters says: "They watch your feet, not your lips!"

How executives act determines how employees react. The true quality of organizational communication is measured by a leader's actions. If you want to assess an organization's communication quality, one need only examine the quality of the individuals who communicate in that organization. When executive decisions and actions concerning safety are consistent, employees, over time, believe that safety is indeed a core value; something they need pay attention to and support. This is called safety leadership, the 'missing element' common to all high loss (and low performing) organizations.

The literature is full of excellent examples of safety leadership, but perhaps one of the best I've heard of involves Zytec Corporation, a Malcolm Baldridge company. Most organizations have a structured and rigorous budget creation and executive review process which starts somewhere around the beginning of the fourth quarter of each year. In many companies, it's more commonly known as the annual 'pad, cut, hack, and trim' process. Zytec has a structured budget approval process, but unlike most other organizations, the safety manager's budget is exempt from review...it's first time final. Talk about sending a message of trust and safety commitment throughout an organization!

Achieving 'corporate intelligence' requires a planned, systematic process for gathering data, and building knowledge in an organization through critical communication systems. Safety communications that work are those that are integrated into the organization, and enabled by leadership which envisions and leads employees to a new communications world called 'TMSIDAK" (Tell Me Something I Don't Already Know)...a strange place to most, but a 'very safe and highly profitable' one for some!

REFERENCES

Albrecht, Karl, The Only Thing That Matters, Harper Collins, New York, 1985

Austin, Nancy and Tom Peters, A Passion for Excellence, Random House, New York, 1985

Barrack, Martin K., How We Communicate - The Most Vital Skill, Glenbridge Publishing Ltd., 1988

Covey, Stephen R., The Seven Habits of Highly Effective People, Simon & Schuster, New York, 1989

Mansdorf, Zack, Communication - The Key to Success, *Occupational Hazards Magazine*, May 1993.

Larkin, Sandar and T.J., Communicating Change - How to Win Employee Support for New Business Directions, McGraw Hill, New York, 1994

Lareau, William, The American Samurai, New Win Publishing Company, 1982

Looking Up - Survey of Workers' compensaiton Issues, *Risk and Insurance Magazine*, November 1994, pp. 30

O'Reilly, Brian, The New Deal - What Companies and Employees Owe One Another, *Fortune,* June 13, 1994, pp. 44-52

Werther, Dr. William B., Dear Boss, What Every Manager Needs to Hear and Every Employee Wants to Say, Meadowbrook Press, New York, 1989

SECTION 11

Measurement
THE STRONGEST FORM OF COMMUNICATING 'WHAT'S REALLY IMPORTANT'!

"What gets measured gets done...but unfortunately,
what gets done often defeats the purpose
of what is being measured."

-Dan Zahlis
"Beware OSHA Statistics"

CAUTION: BEWARE OF OSHA STATISTICS

DAN ZAHLIS

Originally Published as December 1995 cover/feature of' Professional Safety'

"What gets measured, gets done. What gets done may defeat the purpose of what gets measured".

If OSHA recordables are the measure deemed important, organizations will manage categorization of injuries rather than behaviors that lead to them. This approach defeats injury prevention efforts by emphasizing definitions rather than behaviors, and quashes attempts to manage claims with dignity.

A growing number of articles are addressing the value of proactive workers' compensation (WC) claims management techniques. Their messages share a common thread: Treat injured employees with dignity and an organization will ultimately reduce costs. Such methods frequently call for increased initial spending (often exceeding mandated treatment levels) as a means of reducing long-term costs – it costs money to save money.

Unfortunately, the authors of these techniques often fail to identify the primary obstacle faced by those choosing to implement their recommendations; this obstacle is the measurement staple of corporate injury prevention programs: **OSHA statistics used as the primary measure of program effectiveness.** This practice often results in the divorce of two disciplines that should be highly integrated – safety and claims management.

OSHA RATES & INCENTIVES

OSHA statistics are utilized by the agency's compliance officers to calculate injury frequency and severity occurring within industrial facilities. One priority of compliance officers is to calculate these rates using a facility's "Summary of Occupational Injuries and Illnesses," or 200 Log (the lower the rates the better). OSHA also collects recordable data from industries and publishes averages, an exercise that produces corporate 'benchmarking."

As a result, industry has adopted the following logic: If OSHA considers accidents and lost days important, then so should industry. Most businesses take these measurements to the extreme, attaching various incentives to the number of recordable injuries incurred and rewarding low rates.

By assigning incentives to these rates, management recognizes OSHA as the primary customer of injury/cost-reduction efforts. This connection also effectively shifts the burden of control to employees – those theoretically in the best position to control their behavior, yet realistically in the worst position to dictate the speed, manner and priority of their work.

Via incentives, organizations assume that employees will incur fewer injuries in order to obtain established rewards. Without the ability to modify work methods and conditions, however, employees will incur the same number of injuries but report them later and/or less frequently. This approach focuses on the effect (fluctuating OSHA rates), dilutes focus on the root cause of these rates (unsafe behavior) and defeats the ability of claims adjusters to disregard OSHA rates in exchange for proactive claims management.

Using OSHA rates as the primary measurement system almost guarantees reduction in reporting of less severe injuries, which eliminates opportunities for early intervention. Such reductions also eliminate a tremendous body of data on work environment risks and, by default, misdirect management's injury prevention efforts.

Many organizations expend great time and effort managing injury rates by managing injury classification rather than root cause prevention. Generally, this practice – which is much easier than proactive methods – correlates directly with the level of importance that upper management places on OSHA statistics. The net result is:

- 'lower' OSHA rates;

- employees returned to limited duty under most circumstances;

- medical professionals questioned by laypersons;

- attorneys eagerly providing legally mandated services to frustrated employees;

- decreased employee morale

- increase in associated costs.

Thus, OSHA's measurement system is not, holistically, in the best interest of the nation's working class.

PROACTIVE MANAGEMENT

Experienced loss prevention professionals recognize that successful injury and cost-reduction efforts focus on the base of the injury triangle, not the apex (Figure 1). As a rule, reporting must follow these steps:

1. Identify unsafe behaviors.

2. Report close encounters and previously unidentified unsafe conditions.

3. Uncover minor first-aid injuries, moderate injuries ('medical only' recordables) and lost-time accidents.

By increasing an organization's ability to capture the number of unsafe behaviors observed and minor injuries incurred, and taking immediate action toward remediation and medical intervention, severe losses and associated costs will decline. Techniques outlined by proactive claims management experts capitalize on early intervention and open communication. These methods are threatened by nonexistent or late reporting, which is an inherent consequence of assigning incentives to OSHA rates.

Placing priority on OSHA rates results in reporting of injuries that cannot escape reporting – downstream reports based on after-the-fact measurement (which are the most expensive to manage). In addition, fewer reports on less severe injuries will be filed – upstream reports, prior to severe injury (which are the least expensive to manage). Such focus tells employees that reporting should be suppressed until absolutely necessary. Safety professionals should consider whether their locations report triangular data periodically or rectangular data (Figure 2), and the implications of such reporting.

Heinrich's Third Axiom of Industrial Safety states that, in the average case, a person who suffers a disabling injury caused by an unsafe act, has had some 300 narrow escapes from serious injury as a result of committing that very same act. Likewise, a person is exposed to mechanical hazards hundreds of times before s/he suffers injury. Belief in this philosophy necessitates reporting consistent with the triangular format.

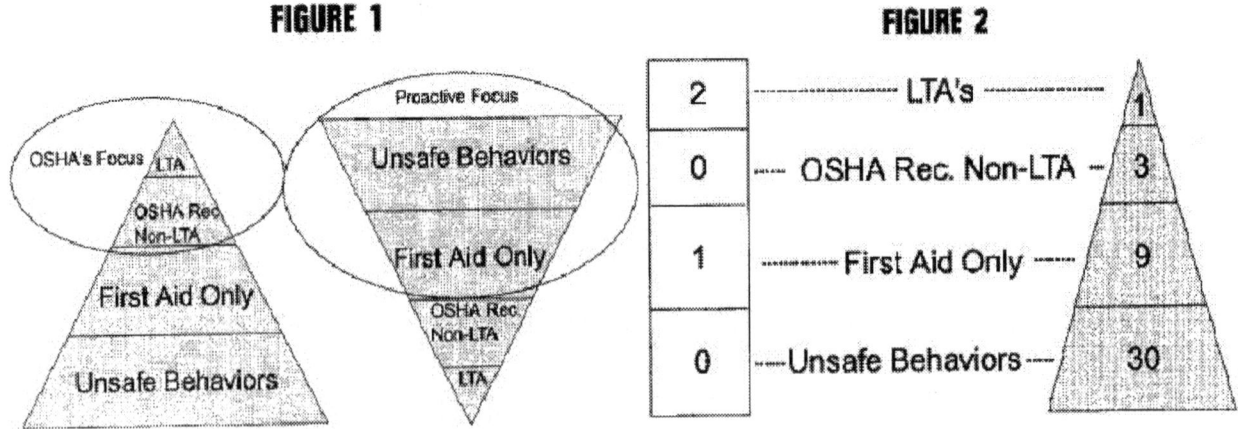

FIGURE 1

FIGURE 2

The irony associated with focusing on OSHA rates is that high OSHA rates can, and sometimes should, be considered positive. When do high rates equate to positive feedback? 1) The practice of sacrificing low OSHA rates in the interest of managing injuries with employees' best interest in mind. 2) Injury prevention programs in their infancy and established programs moving away from focus on the 200 Log. Such programs can ill afford the risk of decreased reporting just as they are "getting their hands around the risks' (increasing the database).

Management should encourage reporting at all stages of severity, with emphasis on unsafe behavior. Management must express its lack of concern over OSHA rates and establish its preoccupation with increasing upstream data available for analysis. Simply stated, the goal should be to increase reporting at all costs – literally. Facilities with young injury-prevention efforts should make clear their goal to have an increase in OSHA recordable rates. Once this practice is well established and reinforced through management action, the art of problem solving and remediation enters in, and OSHA rates will genuinely decline.

A primary goal of claims management is to reduce inherent conflict and eliminate unnecessary third parties, such as medical/legal professionals. Those familiar with WC systems (California's in particular) are well aware of their disparity in costs between litigated and non-litigated claims. Abusive injury compensation systems also fuel dysfunctional relationships between injured employees and their employers. To effectively manage costs in an 'industry unfriendly' compensation system, businesses must create an atmosphere of 'us' (employer and employee) against 'them' (the system).

How does this relate to OSHA statistics an how do such statistics threaten proactive claims management? Claims adjusters and employers need the freedom to treat people with dignity – at the expense of OSHA frequency and severity rates. Yet, how common are the following phrases?

"Jane, go ahead and stay home another day so we're sure you're ready to come back; your benefits will be covered because you've satisfied the waiting period."

"Doctor, I understand that Susie can return to work without restrictions, but it seems that she's still nervous about returning. What do you think about allowing Susie to take off one more day, then work back up to full duty at her own pace, through modified duties?"

"Doctor, if you are more comfortable having John stay home rather than work in a modified-duty position for the next few days, please do. Our primary concern is John's recovery."

"Doctor, if you feel that the prescription medication will be more effective, don't concern yourself with our rates."

"Jane, if you don't agree with the medical/legal report, let's sit down with the doctor and make sure she

237

understands your concerns."

These phrases make many people nervous; to voice these ideas would reap havoc on many organizations' OSHA rates. Such reactions are well-founded in organizations that strive to 'report' lower rates. When fostering good will with injured employees and increasing upstream reporting is the goal, however, such statements become employer/employee team-building and litigation-defeating mechanisms.

Everyone must recognize that fraud and abuse threaten proactive claims management. The key is to apply additional weight to both sides of the scales of justice (proactive claims management and abuse) by dedicating an equal level of relentless effort to eliminating abuse. The most effective means: invite employees to become part of the solution through education and exposure to successful anti-abuse efforts.

BEYOND THE 200 LOG

Table 1 reflects a quarter-to-quarter comparison of two sister manufacturing plants with similar layouts and resources. Presented are the number of claims reported and incurred costs.

Plant 1 employs a traditional focus on OSHA rates, In Plant 2, employees attended an eight-hour training course (before the quarter commenced) that outlined management's desire to increase reporting of accidents and unsafe behaviors – even if the practice increased OSHA rates. In fact, Plant 2's management expressed a desire to increase the total OSHA recordable rate and did not communicate OSHA rates to the workforce throughout the quarter (which had been customary); rates were communicated to Plant 1's employees. To balance the act, Plant 2 also declared war on abuse of its proactive system. During training, management stressed that fraud would not be tolerated and elicited employee cooperation in revealing system abusers.

TABLE 1

Plant	Plant 1	Plant 2
'94 1st Qtr. # of Claims	20	8
'95 1st Qtr. # of Claims	16	27
'94 1st Qtr. Cost	$23,944	$487
'95 1st Qtr. Cost	$47,979	$6,251
'94 Avg. Cost/Claim	$1,197	$61
'95 Avg. Cost/Claim	$2,998	$232

As the data reflect, Plant 1's claim frequency decreased, while claims costs increased substantially. Plant 2's claim frequency increased substantially, while costs increased moderately (as one would expect with an increase in reporting and a reduction of incurred but not reported claims). A simple comparison of average cost/claim tells an interesting story. (Incidentally, Plant 2 is located in California, which has the highest average cost/WC in the U.S. Bulletin No. 94-1; Plant 1 is not).

Plant 2 aggressively manages its claims early (it obtains preliminary data and provides appropriate care); exceeds mandated treatment levels; does not enforce mandated controls (i.e., initial medical control, changes of physicians, etc.); and will not settle disputed claims. The plant invests in eliminating conflict and abuse, a practice that increases initial claim costs and significantly reduces long-term costs.

What the table does not show is Plant 2's substantial database of reported minor injuries, which can be utilized for Pareto analysis and root cause determination. The plant's management and employees also enjoy an improved relationship; they have joined forces to challenge system abuse (claims are currently pending with the local district attorney).

The irony: Plant 1 is considered a 'performer' under the company's measurement system, while Plant 2 is being asked to explain its 'poor' performance. Also ironic is the fact that Plant 2 outperformed 18 sister facilities during an independent audit, yet continues to be identified as a 'below standard' facility.

What is driving the company's performance measurement? Total OSHA recordable rate.

What are the risks? Severe injury and increased costs, and the defeat of an innovative, integrated loss control perspective.

Compliance officers would likely consider Plant 2's program a failure – until they look beyond the 200 Log.

REFERENCES

Bulletin No. 94-1. California Workers' Compensation Institute: San Francisco.

DAN ZAHLIS AND LARRY L. HANSEN

Originally Published as November 2005 cover/feature of' Professional Safety' magazine.

ABSTRACT

Bureau of Labor Statistics(BLS) and National Council of Compensation Insurers(NCCI) annual reports both confirm significant reductions in Incident Rates and Lost time Compensable injuries over the past ten years. Congratulations, Safety Managers! Oh yeah, and by the way...the average cost of medical and indemnity claims continue to escalate, our nation's total cost of' human scrap' continues to mount, Workers' Compensation renewal premiums are increasing, and corporate accrual accounts to fund incurred losses are hemorrhaging in many organizations! Well, that was a short-lived party! The authors propose that this' DISCONNECT'(measures in conflict with results) is pervasive, systemic, and inherently fueled by the traditional measures employed to manage(and manipulate) safety in the business process. The authors examine the harsh reality of the' DISCONNECT' and identify traditional metrics, and the disincentives they create as drivers of under-reporting of incidents, deferral of truth, and ultimate manifestation of increased loss cost. Recommendations, derived from actual experience, are offered to' RE-CONNECT' effective safety (prevention) and loss(mitigation) practices and strategies.

You could feel the excitement in the air. People were giving each other 'thumbs up' 'high fives', and Tiger Woods 'power fists'…"YES!"–'We had done it!'

Safety and Health managers assembled from around the country had just heard the President kick-off our annual EH&S conference with a review of our 2004 statistics…'the numbers'!

"Ladies and gentlemen, I am indeed proud to be here today. You are to be commended for your efforts, your persistence, and your unwavering commitment to 'keeping an eye on the prize'! Due to your leadership, I am pleased to announce that our organization has achieved a level of safety performance never before attained in the history of this company. We have successfully reduced our Incident Rates for a fifth consecutive year, and in doing so have driven our key safety metrics to their lowest levels ever. These accomplishments prove that we indeed have the best programs, the best practices, and the best professionals in the business! I hope the remainder of your conference is productive, and I look forward to the opportunity of coming before you next year with even 'greater news'. Thank you all!"

As 'break chatter' subsided, the meeting chairman called the group back to session and introduced the next senior manager on the agenda. The CFO took the podium and quickly changed everything with this sobering message:

"Ladies and gentlemen, in spite of what you just heard, our Workers' Compensation program is hemorrhaging loss dollars, our accruals have tripled over the past three years, and our loss triangulations are frightful…we must stop the bleeding…FAST! What you're doing isn't working. It's not about 'rates'; it is about 'dollars' –LOSS dollars! Figure out what's wrong; figure out why; figure out a strategy; figure out effective measures; and start addressing the real challenge facing this organization—elimination of loss costs that are diverting revenue from reaching our 'bottom line'. Silence…exit stage right.

This organization had just experienced - THE DISCONNECT!

This, in substance, was the story recited by the Corporate Safety Executive of a large international corporation at a Safety Excellence seminar we recently facilitated. And, even more confirming…when he finished, other participants 'chimed in' with similar stories, experiences, and frustrations confirming the 'DISCONNECT'.

One attendee spoke of the success her facility had achieved in attaining the lowest OSHA Total Incidence Rate (TIR) in their history, a rate of 'point-five' (.5) (some would die for a 'number' that low). But, before the group could offer any commendation on that achievement, she suggested that's exactly what had occurred – the facility had sustained three fatalities that year. The DISCONNECT!

Another participant from the insurance and risk management industry ('Big 3' Brokerage) offered this insight: "Our clients pay us huge amounts of money to help explain and resolve the discrepancy between their OSHA Rates (trending favorably) and their Workers' Compensation costs (skyrocketing), but there just aren't any good metrics out there." The 'DISCONNECT'!

And finally, the Corporate Safety Manager from a sizable military contractor brought discussions to conclusion by exclaiming: "What the heck do they expect us to do about it?" Most definitely -the 'DISCONNECT'!

In many organizations today, workers' compensation costs continue to plague operations and pose a very real threat to corporate profitability. The problems are the same … only now, the answers are different, and so therefore must be the strategies and measures used to combat them. Safety professionals tasked with leading their organizations to higher levels of performance are finding that traditional strategies and prevailing metrics are no longer adequate to meet the challenge. There is indeed a 'DISCONNECT' in safety, and it's driven by 'what is measured' (TIR as scorecard) and its misalignment with 'what is wanted' (cost reduction) in operations.

And this 'DISCONNECT', (loss rates in conflict with costs) is not an anomaly, a temporary situation, or a condition peculiar to an industry segment or unique geography, evidence the 2004 National Council on Compensation Insurance (NCCI) 'SOL Report' analyzing the 'State of the Line' in Workers' Compensation. This report concludes: "Workers' Compensation continues to stand out as a line in need of attention," and further suggests that "not all is doom and gloom". In this regard, it cites the cumulative decline in claims frequency totaling a 39.7% reduction over the past 10 years (NCCI) as a 'positive' development in the line. Although it would be hard to argue that Workers' Compensation is a 'line in need of attention', we disagree adamantly that the continuing decline in claim frequency is a positive development.

A decline in frequency is the 'source' of the 'DISCONNECT'.

Figure 1   **Figure 2**

In spite of the 39.7% decrease in reported claim frequency (Figure 1), the NCCI report further reflects that

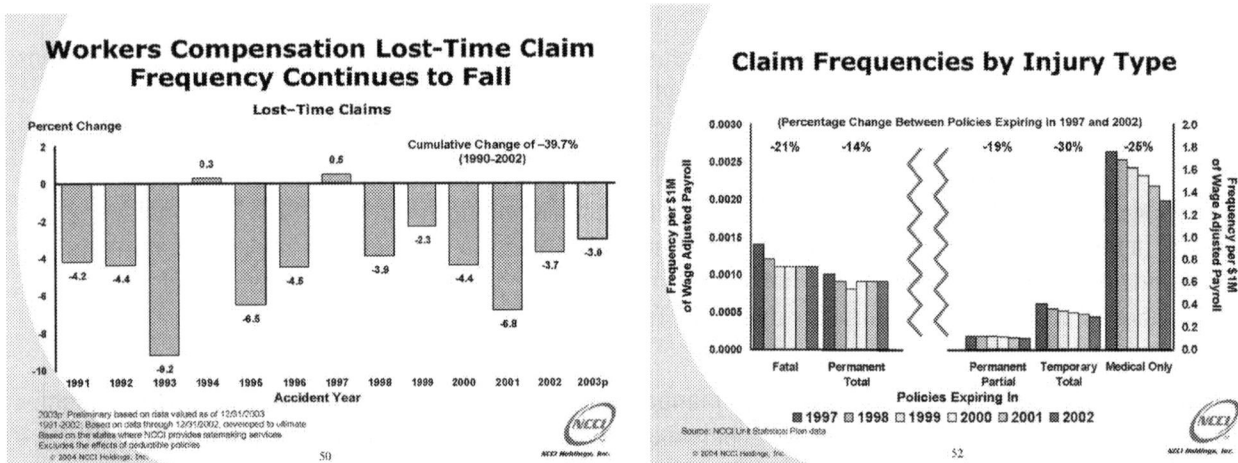

severe claim rates (Fatal, Permanent Total, and Permanent Partial) have remained relatively flat, while lower severity claims (Temporary Total and Medical Only) have been reduced significantly (Figure 2). The analysis further shows that the costs of both the medical and indemnity components of these claims continue to escalate significantly (Figures 3 & 4).

The authors would like to draw the readers' attention to the decline in medical only claim frequency beginning in 1997 (Figure 2) and the steady rate of increase in claims costs depicted over the same period of time (Figures 3 and 4).

These trends although alarming are not surprising to the authors. The escalation of costs in conjunction with a concurrent decline in frequency evidences the 'DISCONNECT' between reporting criteria and effective claims management practices. In our experience, punitive enforcement (based on rates) encourages under reporting of less severe injuries, which leads to 'buried truth and unmanaged exposures' in the work place. These exposures fester as unresolved problems, and ultimately manifest as incidents with higher severity and loss costs. It's the old "pay me (a little) now…or pay me a whole lot more later' dynamic in action!

Figure 3   **Figure 4**

During the past four decades, two management experts, W. Edwards Deming and Peter Drucker, have

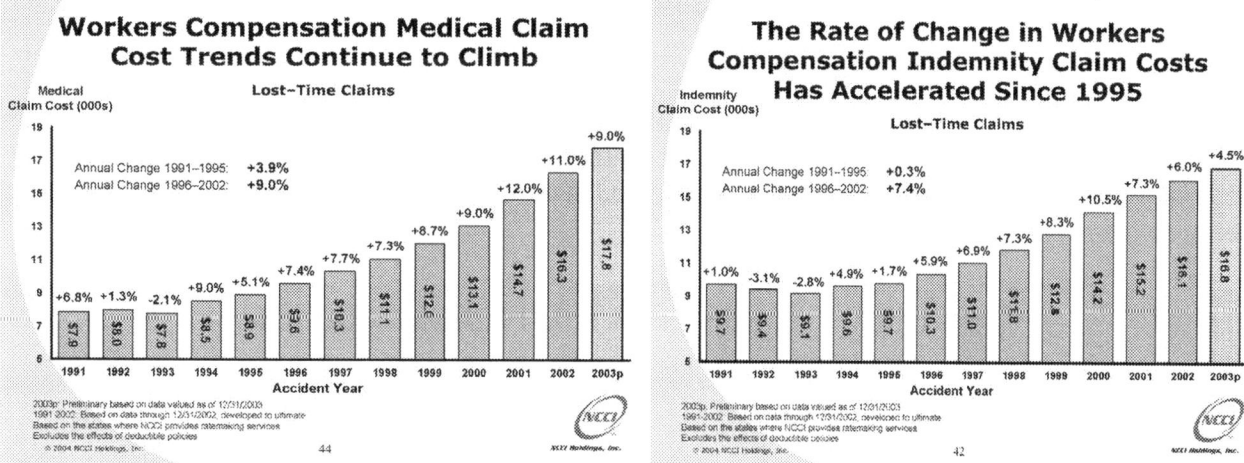

emphasized the critical importance of 'measuring and managing the right things' as requisites for business success. These concepts are equally relevant to safety success as well. Deming identified 'VMO', use of 'Visual Measures Only' as one of the five deadly diseases of American management. He cautioned managers to be wary of, and avoid the dangers of managing by 'the numbers'. He urged managers to seek profound knowledge (the 'reasons' behind the numbers) advising: "Numbers are numbers, numbers are not knowledge."

Peter Drucker, perhaps this planet's most credible management guru, further cautions American managers not to be misguided by a focus on 'the wrong numbers'. Drucker's belief, contrary to that of many managers, is that a management's primary focus should not be on the bottom line, but rather on the critical 'middle lines' of a business. He asserts: "The first duty of business is to survive, and the guiding principle of business economics is not the maximization of profit, it is the avoidance of loss." Loss…also known as cost and expense, are the things that a manager can do something about (manage)…and by doing so, impact the 'bottom line'. It matters not how much revenue (top line) is generated in the business if these 'top dollars' are diverted by loss (middle lines) from reaching margin (the bottom line). Safety (loss/cost) management is not the sole province of the safety manager … it is the shared obligation of every manager in an organization. The safety practitioner's challenge is to grow organizational capability in controlling loss/cost by strengthening line manager competency in measuring and managing the right things at all levels of the organization.

In an October 2003 Occupational Hazards article: "Delivering a 'One-Two' Combination to Flatten the Cost of L.O.S.S.!" the author premised that accidents (alone) don't drive workers' compensation costs – claims do, and that if one doesn't effectively manage both the causes of accidents; 'things'…via an effective safety program, and the causes of claims – 'people', …via effective 'pre and post injury' management practices, all one continues to get are lower loss rates, and skyrocketing loss costs; i.e., the 'DISCONNECT'!

Most agree that the 'Loss Control' function is comprised of two core components; pre-incident prevention (safety), and post incident loss mitigation (claims). Unfortunately, organizations often separate these responsibilities, placing them in Human Resources and Finance, consequentially ending up with competing efforts and metrics; safety gauging success by OSHA incident rates, and claims measuring success by claim cost – the DISCONNECT!

One big question begs for an answer—"Why does safety attach so strongly to OSHA incident rates while most every other organizational function accepts accountability based on cost metrics?" The answers are many (none very good…but many). Following are ten good reasons for safety's continued use of the poor (IR) measures:

TEN 'GOOD' REASONS FOR THE 'POOR' MEASURES

Incident Rate metrics proliferate because:

- Regulators require them.

- The profession tracks them.

- Industry groups compare them.

- Owners let huge contracts based on them.

- Authors cite them.

- Rating bureaus use them.

- Executives believe them.

- Managers are rewarded for them.

- Administrators can manipulate them, and

- It's a whole lot easier than 'performing'!

The authors believe that it's all about the 'Ultimate End Game':

> *"Managers, when forced to manage by the numbers, become highly adept at manipulating the numbers to reward their performance…while true results and purpose fail."*

L. Hansen & D. Zahlis

James Nash of Occupational Hazards has made some notable contributions to this subject including a November 2004 article: 'Weyerhaeuser Fires Plant Safety Manager for RECORDKEEPING ABUSES' and his January 2005 OH online piece entitled: 'OSHA Recordkeeping: Overcoming the Hurdles to Honesty', which begins:

"OSHA recordable injury and illness rates are widely used to assess the effectiveness of workplace safety programs. Yet many observers say the recent five- and six-figure OSHA citations against Weyerhaeuser and General Motors for recordkeeping fraud may just be the tip of an under-reporting iceberg affecting much of U.S. industry."

Within his article, Nash cites corporate safety directors, a corporate medical director, a labor organization leader, safety consultants, a law firm specializing in OSHA compliance, and the Chief of OSHA's division of recordkeeping requirements. Nash's sources agree that attaching positive and/or negative consequences to the reporting of work-related injuries is a harmful business practice.

In this article, Nash also quotes Joe Fortuna, M.D. of Delphi Corporation:

> *I hear people saying, "We have such a low OSHA recordable rate, but our workers' compensation costs keep going up." What's wrong with this picture? – The DISCONNECT!*

Based on the Nash findings, it would be easy to conclude that fewer OSHA log entries are occurring while these unreported OSHA recordables are being captured as compensables on insurance loss runs, thereby driving up costs. However, not the case. Both the Bureau of Labor Statistics, and NCCI report fewer claims.

With 50 years of combined experience in the insurance and risk control industries, the authors concur with Nash's findings - fewer entries are reaching OHSA logs and insurance loss runs due to incentives and consequences that discourage frequency reporting. As a result of this, organizations (and rating bureaus and regulators) become deprived of enormous volumes of 'risk data' (truth) in their operations. Reality suppressed assures fewer opportunities for prevention, and ultimately more severe and costly losses to manifest. Methods known and observed to manipulate recordability include:

- Hiring plant nurses to avoid treatment by physicians.

- Influencing doctor prescriptions.

- Charging employees 'late filing fees' for reporting incidents after occurrence dates.

- Using modified duty programs designed to avoid lost time classifications.

- Reassigning injuries to prior incidents to avoid logging requirements.

- Controlling employee choice of medical provider to retain control over treatments, and

- Using first-aid logs to capture injuries…that often exacerbate due to lack of early intervention.

Manipulating reporting to avoid recordability fuels employee resentment (they're not stupid), resentment fuels HR adversity (desire to get even), adversity fuels abuse (opportunity to get even), abuse fuels litigation (opportunity to get rich), and litigation exposes the company's treasury to the legal system. Employees become conditioned to enter the workers' compensation system as incredulous claimants…more often becoming litigants. And litigation fuels cost!

OSHA recording criteria is not the only reason why claim costs are escalating. Ironically, insurance Experience Modification formulas also discourage reporting. Every injury reported has a direct financial impact on the employer via the multi-year punitive impact on the experience modification rate.

Maybe it's time to 'Think outside the box'…again!

In 1995, Professional Safety Magazine published an article titled "CAUTION: Beware of OSHA Statistics." "CAUTION" was based on actual industry experience that led to quantifiable and reproducible results, premised upon the belief that an inverse relationship exists between incident rates and loss costs; i.e., the higher the incident rate, the lower the cost incurred.

The author of "CAUTION" was hired as divisional Risk Manager for a large global food conglomerate to help correct an operating unit of a larger organization (Plant A). Plant A was experiencing high workers' compensation costs and higher than average OSHA recordable rates. The author's job: 'Fix it'.

An initial assessment determined that Plant A employed no safety professional and that leadership of Plant A had placed safety as a line responsibility with various program responsibilities distributed across many functions in the organization. It was easy to realize that Plant A 'had it right', and had developed a healthy and well-integrated safety culture based on 'partnering'. The author quickly set out to determine the 'drivers' of the high costs and recordable rates.

It was quickly determined that the high costs were due to a lack of local management involvement in the claims administration process, along with under qualified (lacking experience) and disrespectful (disregard for claimant) practices on the part of claim administrators. Claims for this California facility were administered by the corporate headquarters' office in the Midwest. The solution was evident, bring the locus of claims control to the local facility where management could be better informed, advocate on behalf of their injured workers, and insist upon treatments that exceeded legal minimums in order to optimize medical improvement and regain the trust of the employees. This was done.

As for the high OSHA recordable rate? A review of the OSHA 200 Logs revealed that the facility didn't fully understand the criteria for recordability; there were many unnecessary entries. And, this was good!

Over the ensuing months, all employees were trained to understand the process and myths of the workers' compensation system. Employees were openly advised of their rights and entitlements under Workers' Compensation, taught how money moves through the system, how claims costs grow through litigation and forensic medical opinions, and how high costs negatively impact the company. Employees were also taught the dangers of litigation (to the company, their relationships, and their livelihood), and were introduced to free advocates offered by the state of California. In essence, they were taught how to terminate attorneys.

This training uncovered deep-seated employee resentment towards the company based on the incident rate measurement and incentive program imposed by the global parent. As the facility began to focus on 'real issues' rather than 'numbers', and employees began to witness the advocacy of local management on their behalf within the workers' compensation system, they began to open up and identify numerous injuries (and injury causing conditions) that had gone unreported due to pressure and duress associated with the incident rate measure. The truth finally started to emerge; many injuries had been sustained…just not reported.

In response to this employee feedback, all incentives tied to OSHA recordables, and all reports based on 'the dreaded rates' were discontinued. Employees were encouraged to report every incident and near miss immediately. Actually reporting was more than encouraged…it was 'incentivised'…the more Plant A knew, the more they could preemptively manage their risk.

OSHA recordables and claims increased significantly, but a strange phenomenon also occurred: costs began to decrease. Plant A was managing a large book of low severity claims; it was attacking the causes of injuries and was closing claims almost as quickly as they were filed. A plan was born.

One year later? Plant A achieved a 33% increase in reported frequency and a 30% reduction in costs…in the highly volatile CALIFORNIA workers' compensation environment!

A sister facility (Plant B) of the same business unit had developed a more traditional plan consisting of hiring a safety coordinator, centralizing program responsibilities beneath that person, and adhering to traditional practices and OSHA recordable rates. Plant B won a 'Corporate Safety Award' for that undertaking, and was rewarded with a plant wide celebration attended by the usual compliment of visiting corporate dignitaries (at time and expense). Plant B met its goal of reducing its OSHA Recordable rate by 20% - however in the process, increased their loss costs by 80%

Figure 5

And, by the end of the year not only had the 'contra-approach' reduced loss cost by 33%, Plant A had also closed 82% of claims…compared to Plant B's 41% (Figure 6). Plant A had become the second least expensive facility in the company's global network, just behind a small facility in Idaho.

Figure 6

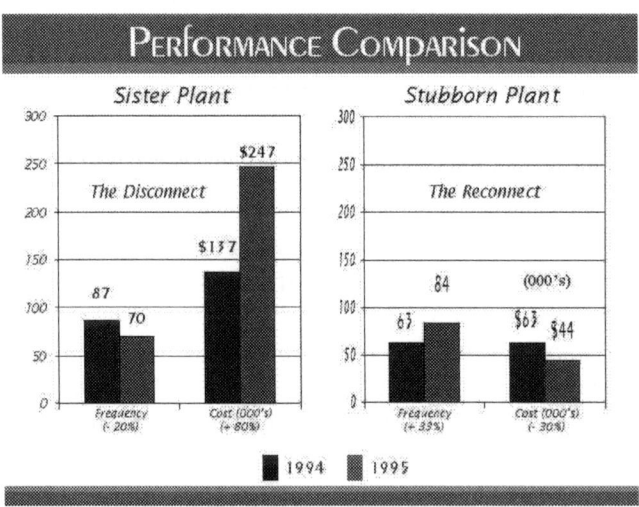

Measurements aimed at reducing frequency are dangerous because of the 'organizational behaviors' they encourage and the 'DISCONNECTS' they create.

RECONNECTING TO HIGH PERFORMANCE

The authors' recommendation is to work toward generating a body of preemptive risk knowledge about your

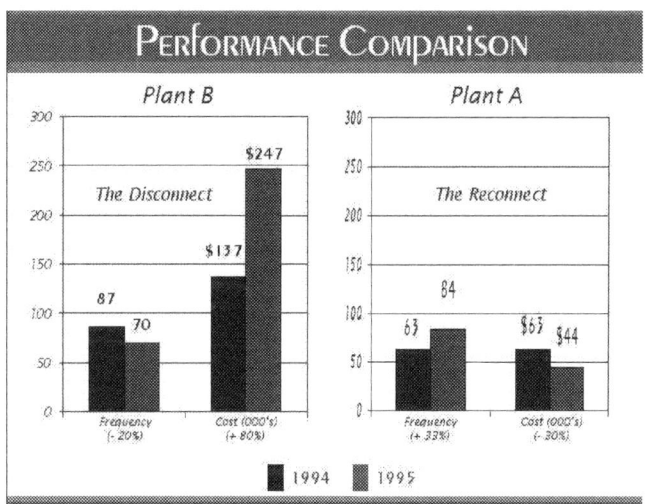

organization; i.e., data on: suggestions; feedback (reinforcing & constructive); hazard reports and abatement status; audit discrepancies and mitigation status; employee satisfaction surveys; incident data such as: nature, source, location, task at time of incident (more is better); employment practice complaints; incentives awarded (based on quantifiable actions!); values threats; supervisor success in motivating more reporting of these metrics; and, any other upstream data about your business culture that guarantees your success by aggressively predicting and preempting failure (cost).

And, eliminate incentives and enforcement based on 'the numbers'. When evaluating safe versus unsafe, consider equations that involve claim cost because such measurements are in sync with other business metrics, and encourage increased reporting of incidents, proactive prevention activities, and good claims management practices directed by local employee advocates. When looking for advocates, recruit employees that grew up in the area and know their coworkers and families personally, irrespective of where they fall on your organizational chart!

For companies interested in 'RE-CONNECTING' safety with cost control, the authors recommend the following steps…in this order:

- Take back control of your claims management process today. Insurance companies do not favor this so you will need to be assertive, and may need to explore the alternative risk financing market. The alternative market consists of self-insurance, high deductible programs, risk retention groups, captives, and other risk financing plans that allow employers to have a more direct impact on the claims management process. The key will be to maintain control over performance by retaining a contractual right to cancel services.

- Locate your claims management administrator within the geographical jurisdiction where your claims will be adjudicated. Identify third party claims administration companies that are located in your area and work towards hiring one of these firms. Local administrators are more knowledgeable about the network of doctors, attorneys, and service providers practicing in your area and they can be much more effective at identifying system abuses.

- Require your carrier or adjuster to learn about your plant, your people, and most importantly your values. Most claims adjusters will be happy to know your employees by their claim number if they're allowed to. Schedule regular visits to your plant and invite injured workers to meet the adjusters during face-to-face discussions. Remind the adjusters that the employees will heavily influence their contract renewal by participating in an annual survey of the adjuster's performance.

- Require adjusters to attend workers' compensation hearings rather than hiring defense attorneys unless absolutely necessary. Defense attorneys are as much a part of the problem as plaintiff (or applicant) attorneys and they both benefit from prolonged disputes. Claims adjusters are very knowledgeable of administrative rules and most hearings are simply casual conversations between the parties in a courthouse lunchroom. If possible, ride with the injured worker to the hearing in the same car.

- Grant the power of advocacy and budget authority to the leaders closest to your people. Insurance companies come and go based on market conditions and rate fluctuations. A company's supervisors, team leaders, and employees will remain employed well after the workers' compensation claim reps and local managers have moved on. Granting authority to resolve workers' compensation claims to local supervision will enhance business continuity and eliminate the disruption (a/k/a distrust) that is inherently associated with cyclical insurance adjusters and management.

- Educate your leaders on the importance of their advocacy and the power of their 'trust-building' behaviors. Provide training to local leadership about the importance of trust in controlling medical costs and reducing litigation. Help local managers to understand that the workers' compensation system is used quite effectively as a 'push back' system and rising costs can be controlled by eliminating the desire to 'push back.'

- Exceed workers' compensation mandates that lead to litigation (i.e. medical controls, benefit delays, etc.). Rather than requiring an injured worker to exceed a three day waiting period to begin receiving indemnity benefits, pay them on day one and save days two, three and four.

- Rather than requiring employees to see the "company doctor,' provide them opportunity to see a doctor they trust and work with the doctor to get the employee back to work when it's in the employee's best interest, even if it costs a few days of indemnity payments to save thousands of dollars in forensic medical bills and attorney's fees.

- Talk with your workforce honestly about their feelings towards OSHA recordkeeping. Hold 'town hall' style meetings to let everyone express their concerns.

- Eliminate measurements and incentives tied to OSHA rates. Eliminate all signs from the workplace that quote the number of hours gone without a Lost Time Accident. Keep your OSHA statistics between yourselves and be prepared to explain the increase. The authors have been through this exercise with OSHA - they quite understand when they observe high frequencies of low severity events in conjunction with effective programs and a positive culture.

- Develop an information system capable of capturing, tracking, and trending copious amounts of operational risk data and converting data to knowledge and knowledge to action. Distribute ownership of the system so that all levels of the organization can enter data directly and browse operational performance. It's amazing what people with differing perspectives can extract from the same set of data.

- Share learned knowledge by posting risk discoveries, dialogues, and activities of every functional type (safety, operations, quality, labor relations, etc.) in all frequented areas, all the time. Never assign a functional title to these postings (e.g. safety report). Keep the information timely and specific to the local operations. The information should be an accurate and truthful reflection of what the culture already knows (e.g. 60% of inspection deficiencies are recorded by supervisor A; or 80% of suggestions are submitted by employees of manager X; or 72% of incentives are awarded to employees of manager Z).

- Educate employees about the workers' compensation system, their entitlements, your approach, and your perspective on the importance of reporting everything. Show employees how dollars flow through the system and how the 'pie' grows when a claim is litigated. Show employees how their 'slice' doesn't grow commensurate with the cost of the claim and ask employees to volunteer their experience. Invite employee advocates from your state to join you in educating your employees. Illustrate how a high frequency of low cost claims is the best approach to prevent more severe losses.

- Conduct regular and frequent claims meetings. Meeting attendees should include the local employee advocate, the injured worker (at worker's option), and the claims adjuster. Don't invite anyone that benefits financially from the loss.

- To get what you want (lower loss cost), measure what you need (better management). Begin tracking average cost per 'anything and everything' that needs to increase to achieve better results; i.e., more reporting (hazards, concerns, complaints, threats and incidents), more communication (postings, flyers, town hall meetings) and more involvement (teams, suggestions, audits) etc. At the very least, use average cost per incident reported (irrespective of severity) because such a metric encourages cohesiveness between prevention and claims management.

Do you have the courage to pull the switch to RECONNECT safety (prevention) with claims management (cost reduction) and create a clear path to the 'higher ground' of excellence?

251

REFERENCES

Deming, W. Edwards, "Quality, Productivity, and Competitive Position", Cambridge, MA, MIT Center for Advanced Engineering, 1982.

Drucker, Peter, F. "Technology Management and Society", New York: Harper & Row Publishers, 1972.

Hansen, Larry L., "Delivering a 'One – Two' Combination to Flatten the Cost of L.O.S.S.", *Occupational Hazards*, October 2003.

Hansen, Larry L., "Twelve 'Unlegal' Ways to Slash Workers' Compensation Costs", *Professional Safety*, June 1997.

Hansen, Larry L. and Zahlis, Daniel F., "Passing an Organizational CAT Scan", *Occupational Health & Safety*, February 2005.

Katz, David M., "How Disney World Keeps Lawyers Out of WC", *National Underwriter*, November 13, 1995.

Mealey, Dennis, National Council of Compensation Insurers, "2004 State of the Line' (Analysis of Workers' compensaiton) presentation, NCCI Web site, March 2004.

Nash, James, "Weyerhaeuser Fires Plant, Safety Managers for Record Keeping Abuses", *Occupational Hazards*, November 2004.

Nash, James. "OSHA Recordkeeping – Overcoming the Hurdles to Honesty", *Occupational Hazards*, January 2005.

Occupational Health & Safety (ohsonline), "Minnesota Reports Comp Claim Rate Down, But Costs Rose.", March 2005.

Zahlis, Daniel F, "CAUTION: Beware OSHA Statistics", *Professional Safety,* December 1995.

Zahlis, Daniel F. "The Hidden Agenda", *Professional Safety*, May 1998.

A UNIVERSAL MODEL FOR SAFETY X-CELLENCE.

SECTION 12

Consequence Delivery

*"Training is the beginning (the first 30%) of learning.
The other 70% comes from the environment,
actions, interactions and consequence systems of the organization."*

~ Bureau of Labor Statistics Report of Findings, 1996

*"The primary tool for unlocking discretionary effort in
an organization is the application of positive
reinforcement for desired behaviors."*

~ Dr. Leslie Wilk-Brasick, Author
"Unlock Behavior; Unleash Profits"

The Universal Model of Safety Excellence

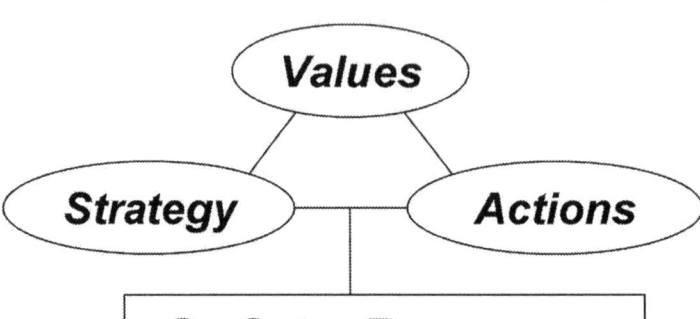

Safety Leadership

Values

Strategy   Actions

Safety Process

Integration

Safety Management

| Human Relations | Communication | Consequences |

A QUESTION OF EXCELLENCE:

CONSEQUENCES
Do we recognize, reinforce and
reward safe performance
equal to all other performance
parameters?

THE 'GOOD REASONS' FOR POOR PERFORMANCE

LARRY L. HANSEN

Perhaps the most frequently asked questions in safety today are: "Why don't employees follow the rules?" -- "Why don't they do what they're supposed to do?" - "Why do they have accidents?" And the most frequently cited answers are: "They're not motivated!" And, "why aren't they motivated?" "Because they don't want to do it." And, "why don't they want to do it?" "Because they're not motivated!"

This 'dizzying' cycle has left managers frustrated and convinced that the primary cause of workplace accidents is PDDT -- (People Doing Dumb Things). Unfortunately, this conclusion is far from correct. To find the real answer to the question: "Why don't employees do what they're supposed to do . . . and have accidents?" - we have to learn to ask better questions.

And that 'better question' is: "Why don't employees do what they're supposed to do and have accidents . . . and the answer isn't that they're not motivated!" When we ask this 'better question," we get 'better answers" to the real causes of our workplace accident problems.

Try this self-test concerning safety in your organization:

What are some of the answers to the 'better question": "Why don't your employees do what they're supposed to do and have accidents (and the answer isn't that they're not motivated) in your organization?"

List answers:

-

-

Your wording might vary slightly, but odds are that your 'better answers' generally fall into twelve specific areas -- The 12 reasons for 'non-performance.' And if we look hard at these reasons, we find that the real causes of employee poor performance involves issues beyond the employee him/her self.

Bottom Line - When it comes to employee nonperformance (why employees don't do what they're supposed to do and have accidents), the causes boil down to two basic issues:

Management:

1.   Did something wrong to or for the employees . . . or

2.   Failed to do something right to or for the employees.

In other words, the basic cause of employee nonperformance is 'poor management!'

Employees don't injure themselves because they're out to get even, or because they're plotting against their companies, or because they're stupid, or because there's a full moon! Employees do what they do (and in some cases get injured) because they are told, directed, or driven to do those things by management policies, procedures, and practices; i.e., because of the boss and his systems.

Let's examine the real reasons for employee nonperformance; i.e., 'the better answers' to the 'better question:' "Why employees do what they do . . . and have accidents."

THE 'TWELVE GOOD REASONS' FOR POOR/ UNSAFE PERFORMANCE.

REASON 1 - They don't know why they should do it.

Here's a pretty typical employee comment: "The boss is a real PIA when it comes to smoking in the lab -- he must be one of those reformed smokers!" - Not! The reason really is fire and explosion from oxygen and flammable gases -- but typically nobody tells employees the reasons behind the rules.

Solution - Develop a comprehensive 'tell 'em why' program - it takes time, but it works. - Reason overcomes ignorance!

REASON 2 -They don't know how to do it.

Most businesses rely upon 'on-the-job training' and as a consequence, most employees don't know how to do their jobs safely -- evidence our national injury rates and the high cost of quality (estimated to exceed 20 percent of revenue) in business today.

Effective training requires time, knowledge and skills that the average employee just doesn't possess; i.e., communication skills, demonstration skills, questioning skills, etc.

Most training efforts are 'designed to fail' — time-based, rather than understanding based; i.e., we need a 'training program,' but it can't be more than 30 minutes . . . that's the 'lunch period'!

In reality, most industry training efforts are nothing more than two people sharing the same space for the same time (generally one hour or less), with one of those individuals required to certify full understanding when the process is complete. And test for comprehension? - No time for that! We simply put them out on the line and let them try to figure it out themselves . . . and we wonder why they make a lot of mistakes trying.

Reasons why employees don't know how:

1.   Managers assume employees already know.

2.   Managers believe that they are teaching when, in fact, they're only telling (passing information not understanding) . . . and

3.   Managers can't waste time teaching when there is so much other 'important work' to do.

Hence, we have the safety failure cycle: hire 'em, place 'em, ignore 'em, blame 'em . . . and pay 'em workers' comp! This creates an economic paradox: employees receive full pay for performance that leads to low quality, damaged goods. late deliveries, excessive waste, and accidents, all of which result in higher costs and lower margin for the employer and tax-free compensation for the employee. Common sense? - No, but is it reality? - Yes! Sometimes economic common sense isn't so common in business.

Solution - Develop a comprehensive 'show 'em how' (training) program. The key elements of which are:

1.   Designate a specific trainer and train that person to teach.

2.   Develop a training manual to standardize training efforts.

3.   Provide reference manuals, materials, and training aids to support training efforts.

4.   Include practice sessions, simulations, and hands-on exercises to make sure that employees get a

chance to make mistakes when it doesn't cost much.

5.   Test for understanding in the classroom and for desired behavior change on the production floor.

REASON 3 - They don't know what to do.

Most businesses keep employees guessing about performance expectations by providing vague job descriptions and unclear responsibilities. Example, 'ASAP' -- does this mean: 'real quick' --- or --- 'whenever you can get to it?' - I personally assume that if someone wants it 'real quick', they'll label it 'BEE: - Before Everything Else!'

The same potential for misinterpretation exists with the typical employee direction: "WORK SAFELY." What does that really mean? (Ask 10 people and you get 10 different answers.)

If you want to improve safety, you must start with well-defined and detailed job responsibilities. Most job descriptions are only a page or two long and, at best, vague! Think about it -- you work 8 hours a day, 5 days a week, 50 weeks a year doing all kinds of 'stuff". Can you compress all that 'stuff' into one page? Hmmm . . . and just what do you do for the remaining 364 days?

Solution - Develop comprehensive job safety responsibilities including specific safety activities; i.e., design safety into job specifications.

Bottom Line - There's a big difference between employee responsibilities and what they do. What you are responsible for and what do you do are usually two very different things. Safety must be clearly defined in both! I'm familiar with a health care organization whose management complains that first line supervisors pay no attention to employee safety -- and they're right. Why? -- Because their first-line supervisory job descriptions contain no responsibilities for employee safety . . . and coincidentally, neither does their performance review process or reward system.

Safety is driven by "what supervisors and employees do' -- issues generally lacking in job descriptions. Job descriptions deal with 'responsibilities' -- not behaviors. Behaviors are what trigger accidents; hence, safe behaviors must be specifically defined, appraised, and rewarded (or not) accordingly.

REASON 4 - They think your way (procedures and rules) won't work.

And guess what -- in many cases, they're right! How often have we heard employees say: "Yeah, looks good on paper but it'll never fly out in the shop?" Typical examples are corporate policies on (lockout/tagout, hazard communication, etc.) OSHA standards. What's put on paper or bound in manuals rarely exists in the shop - that is, unless they're used to propping up an uneven workbench or jamming open a fire door!

Managers in corporate offices write policies and feed those down the organization expecting full implementation. What really happens to these policies? Correct, they're 'filed'; ignored; put on the shelf; or worst of all, 'implemented as written' regardless of the variable conditions in the shop.

The health care industry is a good example. Hospitals and nursing homes are two of the most regulated industries when it comes to safety. They have all the paper, follow all the rules, are heavily inspected, and are generally in pretty complete 'compliance,' yet they are two industries with incident rates far above the national average for total recordable lost time cases and days lost time.

Solution - Implement a comprehensive 'best way' (not necessarily your way) policy development and problem solving process in your organization.

1. Seek employee input when developing programs and policies to encourage ownership and effectiveness.

2. Recognize that it's management's responsibility to 'sell' (convince) not just 'tell' employees what to do.

3. If your way works -- be prepared to defend it with proof. Very few people will argue with facts. -- If you have 'em -- use 'em!

REASON 5 - They think their way is better.

Note - This is not the same as Reason No. 4 (they think your way is wrong). In this case, employees may accept the fact that your way (the rules and the procedures) will work -- but they feel their way is better, faster, easier, or cheaper. A typical safety example is the rule: "Always use guard when operating this machine." And the typical employee interpretation: "If I'm really careful, I can run this job quicker with the guard off and reach in quickly while it's running" -- Zap! -- Minus one hand . . . plus $200,000 in workers' compensation losses!

There's a delicate 'financial' balance between encouraging employees to think; i.e., innovate (find better ways) and requiring them to conform (follow the rules). The real issue is 'good innovation' (i.e., actions for the good of the company) versus bad innovations (actions that can damage the company). Employees must understand the difference . . . and recognize that the bottom line is the financial scale upon which all decisions must be weighed.

Solution - Promote 'good innovation' - thinking that benefits both employees and the company.

1. Ask employees why they think their way is better -- and if it is better, go with it.

261

2. Be prepared to provide convincing information if their way is not better; i.e., show them the results of unintended actions (accidents and their costs, etc.).

REASON 6 -They think something else is more important.

This is frequently known as "I don't have enough time!" If employees claim they don't have enough time, they're probably doing things of a lower priority or in other words, they think something else is more important. We frequently see this situation in safety:

*   Poor housekeeping - "didn't have time to clean"

*   Failure to use right tools (guard) - "couldn't take time to find the right one"

*   Failure to follow care plans, or

*   Failure to use required procedures - "couldn't wait to get help (in health care)"

*   Failure to do safety inspections - "too busy doing 'real work'"

*   Failure to complete accident investigation reports - "no time for useless paperwork"

*   Failure to follow up on safety recommendations - "no time to check it out"

*   Failure to attend/hold safety meetings - "had more important stuff to do"

*   Failure to correct hazards - "can only do one thing at a time"

This list could go on, but as can be seen by just these few examples, the issue at the heart of these problems is 'perceived importance'; i.e., low priority or perceived importance . . . not lack of time.

Managers create confusion by failing to define importance and assign priorities when giving orders or overseeing multiple tasks. Calling everything important only leaves the choices up to employees. The critical difference between successful and unsuccessful organizations is that employees in successful organizations are working on the right things.

Efficiency = Doing things right — being very busy

Effectiveness = Doing the right things — being very productive

Solution - Develop a 'right things' communication and clarification process:

1. Clearly identify priorities on all issues (like safety).

2. Provide employees with definitions or criteria for classifying work that comes to them in a changing environment (help them determine correct priorities).

3. Where priorities frequently change, provide adequate communication to clearly identify changes.

    *   "How bad does the boss want it?" "He wants it 'real bad.'" "That's how he'll get it!"

4. Stop the "it's HOT - ASAP - RUSH" bull -- that's plain management foolishness! Maintaining a constant state of panic in an organization does not solve anything.

REASON 7 - They have no reason to 'wanna do it.'

'Gotta" vs. 'Wanna!" Employees bring two levels of performance to the work environment — required (the minimum level of effort they must give) and discretionary (the variable level of effort they choose to give). Employees choose to do that which they believe will be rewarded.

How many of us have heard something like this: "Six months ago I submitted a suggestion under that new employee suggestion program and I haven't heard a thing since -- guess that safety stuff is just a waste of time!"

Bottom Line - Rewards/recognition influence performance in a positive way. Issues that are ignored become ignored issues. Employee rewards/recognition come from only three sources in the workplace:

1. The work itself - its intrinsic value.

2. Fellow employees - peer recognition

3. The Boss - You!

Question: How many of you like your work — what you do? How many of you like your job? What's the difference?

If the work and coworkers aren't supportive or rewarding -- then it's up to you (the boss).

Solution - Deliver tangible and intangible rewards/recognition (positive consequences) for good performance. People work for pay -- they perform for rewards. If you don't compliment and reward the performance you want, you won't get the performance you need even though you're paying for it.

REASON 8 - They think they are doing it.

One of the core reasons for nonperformance is inadequate feedback -- employees that are uninformed or misinformed about their performance.

Question - "How do employees know when or if they're doing a good job?"

Better Question - "How do you know when or if you're doing a good job?" Think about it -- is it 'really' because your boss tells you so? Most organizations, unfortunately, subscribe to the "no news is good news" school.

Ken Blanchard calls this practice the "Leave-alone — ZAP!" i.e., managers say nothing until there's a screw-up and then they nail the employee.

When we ask employees, here are some typical answers we get:

1. "When the boss leaves me alone, I know I'm doing okay";

2. "We must be doing all right, we haven't got any safety speeches for a while";

3. "It must be okay, "they" complain when there are too many accidents";

4. "When we don't get our reports back, we know we're right";

5. "When the boss doesn't glare at us in meetings, we know we're doing a decent job";

6. "As long as they keep harping on production, we know we don't have to worry about safety" . . . and ` the biggie:

7. "As long as the paycheck keeps coming, we know we're still employed!"

Lack of feedback (non-information) is what employees typically get and interpret as good performance.

Solutions - Implement timely, positive feedback systems as close to the job as possible.

Postings, newsletters, signs and notices, memorandums, group meetings, open forum question and answer sessions, one-on-one contacts, and short notes of approval.

Continuous feedback influences performance continually. Periodic feedback impacts performance only at specific intervals - feedback is most effective on a frequent basis.

One automotive assembly company posts monthly production and quality quotas on whiteboards at key points along each process line and updates these on an hourly basis. There are no safety goals. Guess what their performance in production and quality is. Now guess what their performance in safety is. And, do you know they can understand why their WC loss experience is so bad (EM = 2.62); a/k/a the high cost of low communications.

REASON 9 - They are rewarded for not doing it.

Here's a real life example -- observed in a fabrication plant:

While walking through the plant, a supervisor spotted metal banding on the floor and picked it up instead of making the employee who put it there pick it up. Who's training whom? -- Hey, maybe that's the answer, put employees in charge of supervisory training!

Solution - Make employees correct their own safety hazards.

REASON 10 - They are punished for doing it right!

This is also known as 'shooting the messenger.'

Some examples of punishing good performance in the workplace are:

- "The reward for good work is _____!" Employees who perform difficult tasks well are assigned more difficult tasks.

- In the military, it's called "Never volunteer."

- Employees who make good suggestions are assigned to carrying out the extra work created by those suggestions.

- The death of a 'good idea.' Employees that try to innovate are told - "why can't you just follow instructions."

The unfortunate reality of small minds in high places; i.e., bosses build themselves up by cutting others down. An employee making a special attempt to cleanup his work area is told by his supervisor - "You finally have found something that fits your qualifications."

Bottom Line - People avoid ridicule -- punish them for what they're supposed to do and they'll stop doing it.

Ever wonder why hardly anyone at seminars ever raises their hand when a presenter asks, "Any questions?"

Solution - Remove intentional (and unintentional) punishment for good performance.

- When employees come to you for help - give them help, not criticism.

- Eliminate all sarcasm concerning good performance.

- Example -- In an aluminum smelting company, managers and supervisors refer to employees on RTW/modified duty as 'Wusses!'

- Look at the need for procedures/policies from an employee majority point of view - most corporate policies are based on the distrust or need to control a few, not the many. Guess what message 'the many' receive (Example - 100 percent eye protection).

- Other examples? Dress codes, restroom breaks, etc.

REASON 11 - They perceive no negative consequences for not doing it.

Safety is just a matter of doing what's important. When supervisors don't submit monthly safety inspections, or don't complete accident investigation reports, or when employees repeatedly fail to follow safety rules or wear safety equipment — and nothing happens, safety just isn't important.

Is keeping an OSHA log important? At one large construction company it is; you'll be fired if it's not up-do-date.

If management does nothing about non-performance, guess what happens? - It continues.

If we 'cheat the system' and 'beat the system,' what develops?...Bad habits.

Here's a real (and unfortunately typical) problem - a plant manager consis¬tently fails to support a safety program and enforce required compliance initiatives gets into a dispute with the safety manager. The vice-president tells the safety manager to 'back off' because the plant is making a profit. That organization has just encouraged nonperformance via lack of negative consequences.

Another example - a lab technician is repeatedly reprimanded for violating the company's policy requiring safety glasses in the lab. After a supervisor's reprimand, the employee returns to the lab wearing glasses and the supervisor, pleased with this change of heart, offers a compliment -- to which the employee wiggles his fingers through the empty frames. It's interesting what employees will do when there are no negative consequences for nonperformance.

Solution - Assure that your organization has (and applies) negative consequences for nonperformance.

REASON 12 - They can't overcome obstacles beyond their control. In many cases, employee nonperformance is due to physical, procedural, or organizational barriers; such as:

1. Inadequate resources (including staffing) - this is a typical problem in the health care industry. Safe procedures require two persons to lift residents and patients. The reality in these facilities is that often only one person is available; i.e., short staffing.

2. Conflicting messages - management typically tells employees that 'safety is job one' (i.e., 'safety first') -- but when push comes to shove, it is clear that production quota is what it's all about. (Forget that 'safety fluff!')

3. Authority not delegated with responsibility - this is a typical dilemma faced by chairman of safety committees -- management says it's your job to 'do safety' . . . but don't interfere with the real work that we do around here (production).

What are some other obstacles beyond the control of employees? Equipment malfunctions, inadequate purchase specifications, poor maintenance, etc., etc., etc.

Solution - To improve employee performance from obstacles beyond their control, managers must be willing to stand accountable and plan, improve, repair and/or remove obstacles which impede good performance.

Managers must be willing to 'fix the process, not the blame.' They must recognize that most employee excuses for nonperformance aren't due to bad attitudes but to poor management, planning, policy and procedure.

If you can't remove the obstacle, then you must provide employees with a strategy for overcoming it. And if you can't give employees a strategy for overcoming it, then you better give them something else to do because they're not going to do it because 'no one can do it.'

The real answer to improving employee performance is to remove the causes of nonperformance using the concept of 'preventive management.' This concept is similar to that of 'preventive maintenance;' i.e., planned interventions to maintain 'peak' mechanical efficiency; i.e., keep things running right. The concept of preventive management is 'planned intervention to maintain human efficiency' -- keep things running right on a human level.

Question: Does good management occur before or after employee performance? – YES! Good management is preventive intervention (actions before) and consistent follow through (actions after)!

For managers to be effective, management attitudes, philosophies, and policies must be converted into management 'actions" (behaviors) that influence people's performance. Or, as Scott Geller puts it: "Managers must act people into thinking safely."

Applying management as an intervention means that managers must do specific things at specific times to influence the eventual outcome of their people's performance. Preventive management overcomes the 12 reasons for employee nonperformance!

BEFORE THE WORK BEGINS

B-1 - "EXPLAIN WHY" - TO ELIMINATE: THEY DON'T KNOW WHY THEY SHOULD DO IT.

B-2 - "TRAIN, DEMONSTRATE, AND TEST"- TO ELIMINATE" - THEY DON'T KNOW HOW TO DO IT.

B-3 - "DEFINE JOB BEHAVIORS" - TO ELIMINATE: THEY DON'T KNOW WHAT TO DO.

B-4 - "SEEK EMPLOYEE INPUT" - TO ELIMINATE: THEY THINK YOUR WAY WON'T WORK.

B-5 - "OPENLY COMMUNICATE" - TO ELIMINATE: THEY THINK THEIR WAY IS BETTER.

B-6 - "FIX CLEAR PRIORITIES" - TO ELIMINATE: THEY THINK SOMETHING ELSE IS MORE IMPORTANT.

AFTER THE WORK COMMENCES

A-7 - "REWARD GOOD PERFORMANCE" - TO ELIMINATE: "THEY HAVE NO REASON TO 'WANNA' DO IT."

A-8 - "PROVIDE TIMELY FEEDBACK" - TO ELIMINATE: THEY THINK THEY ARE DOING IT.

A-9 - "DON'T IGNORE NON-PERFORMANCE" - TO ELIMINATE: THEY ARE REWARDED FOR NOT DOING IT.

A-10- "DON'T CRITICIZE GOOD BEHAVIOR" - TO ELIMINATE: THEY ARE PUNISHED FOR DOING IT RIGHT.

A-11- "CONFRONT NON-PERFORMANCE IMMEDIATELY" - TO ELIMINATE: THEY PERCEIVE NO NEGATIVE CONSEQUENCE FOR NOT DOING IT.

A-12- "DO YOUR PART" - TO ELIMINATE: OBSTACLES BEYOND THEIR CONTROL.

Source - Why Employees Don't Do What They're Supposed to Do and What to Do About It, Ferdinand Fournies (Liberty Hall Press/Tab Books, Blue Ridge Summit, Pennsylvania, 1988).

HIGH PERFORMANCE REQUIRES DUAL STRATEGIES

LARRY L. HANSEN

Its' not about making employees happy…it's all about making employees 'productive'…which just happens to require that they be 'happy'.

In his recent article: "Happy Employees Don't Equal Happy Customers", Jim Harrington of Ernst & Young highlights the common "management misconception' that employee satisfaction and high work performance are driven by the same factors. His article clearly indicates that employee satisfaction and work performance are outcomes of very different organizational issues. Harrington has effectively surfaced what Tom Peters refers to in this case as "a blinding flash of the obvious" … the strategies which create happy employees don't necessarily equate to those which create happy customers!

Despite the clarity of the literature, many managers continue to believe and manage as if there's a direct linkage. Victor Vroom's writing, "Work and Motivation" best differentiates these two issues and the strategies that impact each. Although there are commonalties, there are far more differences and contrasts in these strategies and these are the key points managers must understand.

Vroom contends that 'satisfiers' (what makes employees happy) in the workplace are far different from 'motivators' (things which create employee desire to perform). His conclusion:

> *"There seems to be little correlation between employee satisfaction and organizational performance."*

- Vroom

The corresponding message for practicing managers is: "Employee satisfaction and work performance are two very separate issues which require two very different strategies and practices.

Vroom suggests a performance formula, which links ability with motivation -- it is:

$$Performance = (ability \times motivation)$$

$$\overline{\qquad\qquad\qquad}$$

$$(Can\text{-}Do)\ (Want\text{-}To)$$

In this formula, employee satisfaction (like-to) has no identified role.

Although there is little 'direct correlation' between employee satisfaction and high work performance, there is some 'commonality of issues' that relates these two factors.

Vroom identifies (Table I) both the common and differentiating strategies and management practices which impact satisfaction and performance respectively.

Table I

What Increases Employee Job Satisfaction	$\neq$	What Increases Employee Job Performance

"Satisfiers"	**"Motivators"**
• Perception of equitable wages	• Perception of overcompensation
• Social environment (interaction)	• Group 'cohesiveness" (strength of beliefs)
• Considerate/participative supervision	• Considerate/supportive supervision
• Individual personality	• Participation in decision making
• Promotional opportunities	• Specialization of work
• Expected level of rewards	• Feedback:
• Work hours/schedules	• Learning
• Control over work/pace	• Performance
• Fair treatment	• Results
• Opportunity to use skills	• Abilities are valued
	• Opportunity to achieve success

271

SATISFIERS VS. MOTIVATORS
A/K/A: "DIFFERENT STROKES FOR DIFFERENT FOLKS"

With a clear understanding that employee satisfaction and work performance are driven by issues that share common traits but require different approaches, managers hence need design strategies and tactics which 'influence both.'

The following discusses these issues and offers comment on commonality and key differentiations, which require that peak performance, be a delicate 'mix and match' process.

Management of Satisfiers - Job satisfaction increases when employees perceive:

S1. **Wages to be equitable** - Employees are satisfied when they perceive their wage to be fair and equitable on a relative basis . . . a fair day's work for a fair day's pay!

   • **Strategy** - Develop and administer compensation programs (wage scales, incentive plans and administrative policies) which create the perception of fairness and equity in pay practices.

S2. **An open social environment** - Employees have high satisfaction when they have an opportunity to interact and communicate with others in the work environment.

   • **Strategy** - Provide workstation design and workplace configurations which allow employees to interact and have a 'direct line of sight' contact with other employees.

S3. **Supervisors as supporting and considerate** - Employee satisfaction is high when they perceive their supervisors to be helpful.

   • **Strategy** - Define and design supervisory roles to be facilitative and supportive; i.e., a focus on coaching rather than controlling.

S4. **Opportunity to express individualism** - Employee satisfaction is increased when they have an opportunity to express individuality in their work.

   • Strategy - Provide jobs which allow employees flexibility and choice in work methods. Allow them to demonstrate their individual craftsmanship.

272

S5. **Promotional opportunities** - Employee satisfaction is increased when they perceive opportunities for advancement.

- Strategy - Design performance management systems which clearly link work performance to growth and advancement opportunities. Appraisal and evaluation processes should be designed to assure that high performers are recognized and moved forward in the organization.

S6. **Rewards equal to expectations** - Employees are satisfied when they are able to achieve the rewards they perceive to be commensurate to the efforts they expend.

- **Strategy** - Design and administer compensation and reward systems which are in line with fair market values and realistic to employee expectations. Employees know the fair market 'worth' of their efforts -- make sure your systems conform and allow them to achieve this.

S7. **Reasonable work hours and schedules** - Employees are satisfied when work hours and schedules meet their personal needs.

- Strategy - Provide work hours, schedules and 'time off' benefit packages, which are flexible to individual needs. When it comes to benefit packages -- 'One size fits all' -- generally doesn't!

S8. **Opportunity to control work methods and pace** - Employee satisfaction is increased when employees feel they have 'control over their work."

- **Strategy** - Design work flows, schedules and processes which allow employees maximum control over their work. Focus on the end results more than the means employed to accomplish them. Move their minds, not their feet!

S9. **Fair treatment** - Employee satisfaction increases when employees believe they are being treated fairly and justly.

- **Strategy** - Design and administer work assignments, performance measurements, and 'checks and balance systems' which ensure that work is allocated, measured and rewarded fairly. Avoid 'favoritism' at all costs -- favoritism kills both spirit and performance in the workplace.

S10. **Opportunities to use individual skills** - Employee satisfaction increases when employees feel they have the opportunity to use their unique abilities.

- **Strategy** - Design work assignments and/or job placements to maximize the use of unique employee skill sets. Allow employees to build upon their unique strengths and contributions.

MANAGEMENT OF PERFORMANCE MOTIVATORS
JOB PERFORMANCE INCREASES WHEN EMPLOYEES PERCEIVE:

P1.   **A high level of compensation** - Employees tend to work harder when they believe they are receiving high compensation for their efforts as compared to others.

- Strategy - Design compensation structures, administrative policies, and communications which ensure that employees perceive their wage and benefit packages to be greater than industry comparable.

P2.   **Group cohesiveness** - Employee work performance increases when they are allowed to forge work groups with strong values and beliefs.

- Strategy - Develop and support team-based organizations to ensure that group cohesiveness remains 'positive' and that the strength of the group compliments the corporation's mission.

P3.   **Considerate and supportive supervisors** - Employee work performance improves when they believe their supervisors are supportive of their efforts.

- Strategy - Assure that supervisory roles are defined and implemented supportive and mentoring rather than controlling and authoritarian.

P4.   **Opportunity to participate in decision making** - work performance improves when employees have the opportunity to provide input on decisions, which will affect them.

- Strategy - Encourage employee involvement and participation in organizational structure and policy.

P5.   **Specialization of work** - Work performance increases when work is 'specialized.'

- Strategy - Develop work systems, procedures and job placements that maximize the use of 'specialized employee skills.'

P6.   **Adequate feedback** - Work performance improves when employees receive adequate and frequent feedback concerning their work.

- Strategy - Develop and implement progressive communication and positive reinforcement systems to keep employees informed of their performance and results.

P7. **Their abilities are valued** - Work performance increases when employees believe they are treated as if their abilities have high value and make a difference to the organization.

- Strategy - Ensure that performance evaluation processes reward unique employee abilities and contributions.

P8. **Opportunity to achieve** - Employee work performance increases when they perceive an opportunity to 'achieve.' Employees want to be winners too!

- Strategy - Develop job designs and placements, which afford employees an opportunity to maximize their special skills and succeed in their undertakings.

Conclusion - It's management's job to "get things done through people.' This, however, is not an easy task. It requires a correct 'mix' of strategies to attain both motivation and satisfaction. A peak performance organization fully understands the differences, which maximizes both performance and job satisfaction in the organization.

REFERENCES

Vroom, Victor H., Work and Motivation, John Wiley & Sons Incorporated, New York, New York

WHY EMPLOYEES HAVE ACCIDENTS... AND WHAT YOU CAN DO ABOUT IT

Two questions come to mind when thinking about employees and their safety on the job:

1. "Do people, like animals, have an innate instinct toward self-preservation?" In other words, will people naturally engage in behaviors that protect their own safety?

Research in the animal kingdom shows that it is virtually unheard of for animals to carelessly risk their own lives. However, it seems that, in many cases, human beings will engage in such behaviors. The extent to which they will do so varies from person to person, however. While some people take steps to live safely most of the time, others seem to be willing, for various reasons, to compromise their own safety.

For example, many people drive without seatbelts, drive at reckless speeds, drink while driving, use illegal drugs, smoke cigarettes, live on diets of junk food, fail to exercise, etc. These are all behaviors they know will place their safety at risk, yet they engage in them anyway.

2. "Will those who do have an instinct toward self-preservation short-circuit that instinct under some circumstances?"

Larry L. Hansen, CSP, ARM — an organizational performance consultant with L2H Speaking of Safety (Baldwinsville, New York) — and Kelly Hansen — safety manager for Mosinee Paper Co. (Mosinee, Wisconsin) — are two safety professionals who have observed the dynamics of at-risk behavior and believe it has systemic as well as individual causes.

They believe that employers do a good job of screening out applicants who lack skills or demonstrate tendencies for high 'at risk' behaviors. Consequently, the large majority of workplaces tend to be staffed with people who want to work safely (those who are covered in Question 1).

However, over time, because of workplace values, philosophies, policies, and practices, these safety-minded people will often be willing to short-circuit their natural safety instincts for a number of different reasons (the issue addressed in Question 2).

What are the things that can override a person's natural instinct to behave safely? The Hansens point to a number of drivers, the four most powerful being ...

• Recognition and compensation systems,

• Management values and expectations,

• Fear, and

• Lack of knowledge/skills.

In other words, say the Hansens, employees will be willing to put their own personal safety at risk if the compensation system rewards unsafe behavior, if their employers expect or allow them to work unsafely, if they are afraid to work safely, and/or if they lack knowledge of what is safe.

Using these four classifications, the Hansens have come up with 10 reasons why employees who engage in safe behaviors off the job are willing to engage in unsafe behaviors on the job.

RECOGNITION AND COMPENSATION SYSTEMS

1. All too often, employees are rewarded for unsafe behavior. Formal rewards include pay for performance (bonus plans, incentives, piece-rate systems), which encourages shortcuts and increased risk taking. Informal rewards include verbal recognition for extra productivity at the expense of safety.

    This is a particularly interesting concept — that employees are willing to compromise their natural instinct to engage in safe behavior in return for money or for verbal recognition.

MANAGEMENT VALUES AND EXPECTATIONS

2. They can get away with unsafe behavior. This involves a lack of consequences.

    "All behavior is controlled by consequences," explains Kelly Hansen. "When there are no consequences, either positive or negative — or when consequences are insignificant and delayed — employees will resort to 'quick and easy,' which often equates to 'unsafe.'"

    This is also an interesting concept — that employees will surrender their personal instincts to work safely and, instead, make risk-taking decisions based on whether or not their employers offer consequences.

3. The organizational culture encourages unsafe behavior. According to Larry Hansen, every organization has two safety processes. One is the formal safety program, which includes the written rules. The other is the more powerful safety culture, which includes the unwritten 'rules' and which ultimately determines what really happens.

    "All too often, the safety program is compromised by the culture," says Hansen.

4. Safe performance is ignored. When employees put forth the effort to do a task safely, and supervisors ignore it or fail to positively acknowledge and reinforce it, employees conclude that safety is not all that important, says Kelly Hansen. "They then revert to practices that are the quickest and easiest."

5. They try to do the right thing in a wrong system. "Many injuries are sustained when employees are exposed to process-inherent hazards and react spontaneously to prevent injury to people or damage to products," says Larry Hansen.

    Example: A stack of products begins to fall, and the employee tries to prevent the accident, injuring himself or herself in the process.

6. Risk is designed into the task. "Despite the abundance of information available on automation, job modification, and process redesign, many organizations continue to operate under the regressive belief that it's cheaper to bend and break backs than it is to bend steel," continues Larry Hansen.

    "As long as employers are willing to finance the 'cost of loss' incurred from doing business unsafely, injuries from repetitive motion, lifting, twisting, and reaching will continue to produce soft tissue, nerve, and muscle injuries."

7. The measurement system demands it. Many managers believe that their primary objective is production (pushing volume and getting units shipped). But Larry Hansen believes the proper objective is productivity, which involves maximizing output (production) while minimizing input (resources consumed).

    "You must manage both output and input effectively if you are to optimize profitability," he states. "The question is: Do you want to make more money or to keep more of the money you make? To achieve the latter, you need to control loss."

FEAR

8. They're punished for working safely. In work cultures where volume, quotas, and numbers are the high priority, employees know the consequences of failing to make their numbers. "Given the option of

complying with safe procedures or making their quotas, employees know that there is no option," he suggests.

9. They fear for their jobs. Recessionary influences emphasize cost-cutting, sending a clear message to employees that "If you can't get it done, someone else will."

"These unfortunate workplace realities increase the level of fear and encourage greater levels of risk-taking to meet elevated demands," he adds.

LACK OF KNOWLEDGE/SKILLS

10. They believe they are working safely. "In many organizations, minimal effort is devoted to the sort of employee orientation, safety training, and ongoing supervisory follow-up necessary to ensure that employees understand safety requirements and are applying safe procedures on the job," says Larry Hansen.

"Most safety programs are set up to meet minimum regulatory requirements, not to ensure that employees work safely. As a result, employers get marginal results at best."

He cites the following example: "I receive calls from employers who are looking for technical/regulatory safety training for their employees. The training issues are detailed and complex, but often the employers say that time is limited and they can allot only 45 minutes for the training — or that it must take place between shifts." Without realizing it, they are making their priorities crystal clear.

PERSPECTIVE

While virtually everyone would agree that employees will work unsafely if they are unaware that they are working unsafely (lack of knowledge), and many would agree that employees might work unsafely if they are afraid not to (fear), the Hansens raise two other interesting ideas:

• employees will agree to work unsafely in return for rewards (compensation system), and

• employees will agree to work unsafely because other people directly or indirectly expect them to (management expectations).

"If you see employees as the problem in safety, your facility's safety performance will never improve," says Larry Hansen. "According to W. Edwards Deming, 96 percent of the defective outcomes of a system are due to the process you design, not the individuals in the process."

To improve safety, the Hansens believe that you must reorganize your business systems, leadership values, and management practices to encourage and reward safe practices, rather than relying only on your employees to voluntarily utilize their natural instincts to behave safely.

REFERENCES

"This article is republished here with the permission from *Safety Management,* Issue #462, Sept. 2001, Copyright 2001, Aspen Publishers, Inc. All Rights Reserved. For more information on this or any other Aspen Publications, please visit http://www.aspenpublishers.com"

SM #462-03 (Employee Safety, pages 1, 2 – 3)

Interviewer: (2) William Atkinson (PO Box 38, Carterville, IL 62918) Interviewee: (5) Larry L. Hansen, CSP, ARM, organizational performance consultant, L2H Speaking of Safety (PO Box 532, Baldwinsville, NY 13027; PH: 315-383-3801; EM: llhsos@dreamscape.com)

SECTION 13

PROCESS
Total Quality Management (TQM)

"Most of the problems in organizations today exist because of management. Their thinking is seriously flawed; it focuses on authority, legislation, and control — not people, values, and quality outcomes."

~ W. Edwards Deming
Cited by Karl Albrecht in "The Only Thing that Matters"

"People, however different, when placed in the same system, tend to produce similar results"

~ Peter Senge

SAFE is but one outcome of the PROCESS called management.

The Universal Model of Safety Excellence

Safety Leadership

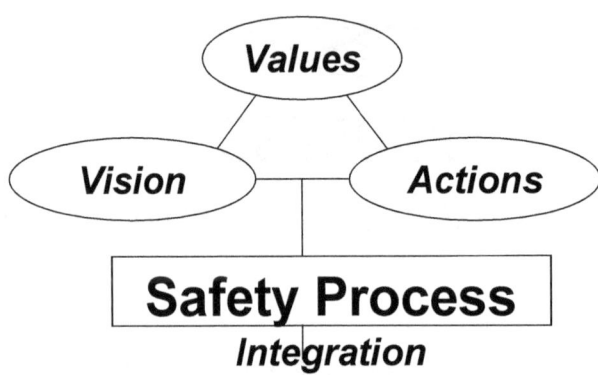

Values

Vision   *Actions*

Safety Process

Integration

Safety Management

Human Relations	Communication & Measurement	Consequences

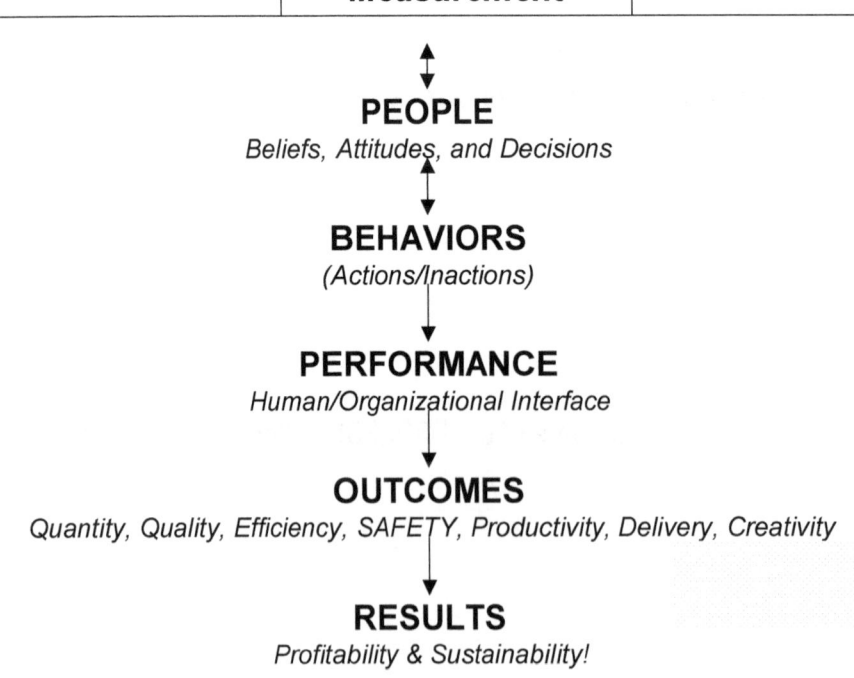

PEOPLE
Beliefs, Attitudes, and Decisions

BEHAVIORS
(Actions/Inactions)

PERFORMANCE
Human/Organizational Interface

OUTCOMES
Quantity, Quality, Efficiency, SAFETY, Productivity, Delivery, Creativity

RESULTS
Profitability & Sustainability!

APPLYING DEMING'S PRINCIPLES TO SAFETY MANAGEMENT

~

"IN 50 WORDS OR LESS…"
(44 TO BE EXACT)

A transition paper identifying the correlations and relevance of Dr. Deming's Total Quality Management principles to improving "*SAFETY*" in the workplace.

Transitions and comments by
Larry L. Hansen, CSP, ARM

In the 1970's and 1980's, I became a firm believer that Total Quality Management (TQM) principles held tremendous opportunity for improving our efforts to manage *safety* in the workplace. The more I read of Joe Juran, Philip Crosby, and most of all W. Edwards Deming, the more I became totally convinced that their principles were correct and could be directly applied to managing *safety* in the workplace, i.e., Total *Safety* Management. In my readings on TQM, I came to realize that TQM wasn't about quality…it was all about 'MANAGEMENT'. Ultimately, I read Dr. Deming's work entitled: "Quality, Productivity & Competitive Position", and as I read the following section of that book, it became clear to me that Deming's philosophies indeed could (and should) be applied to *Safety* Management. By transposing but a few words (from a quality context/vocabulary to *safety*), one can readily see how he identifies and addresses those critical issues most common to under performing *safety* organizations.

A Report to Management
by W. Edwards Deming

"This report is written at your request after study of some of the problems that you are having with low production, high costs and *worker injuries* (1) which altogether, as I understood you, have been the cause of considerable worry to you about your competitive position.

My opening point is that no permanent impact has ever been accomplished in improvement of *safety* (2) unless the top management carries out their responsibilities. These responsibilities never cease; they continue forever. No short cut has been discovered. Failure of your own management to accept and act on their responsibilities for *safety* (3) is, in my opinion, the prime cause of your trouble, as further paragraphs will indicate in more detail.

You assured me, when I began this engagement, that you have a *safety **program*** (4) in your company. I have had a chance to see some of it. What you have in your company, as I see it, is not a ***safety program*** (5), but guerrilla sniping – no organized system, no provision nor appreciation for control of ***accidents*** (6) as a system. You have been running along with a fire department that hopes to arrive in time to keep fires from spreading. Your **safety** (7) department has done their duty, as I understand it, if they discover a **hazard** (8), which might cause an ***accident*** (9) (even legal action) if it went ***uncorrected*** (10). This is important, but my advice is to build a system of **safety** (11) that will reduce the number of fires in the first place. You spend money on **safety** (12), but ineffectively.

You have a slogan, posted everywhere, urging everyone to work **safety** (13), nothing else. I wonder how anyone could live up to it. By every man *working **safer*** (14)? How can he, when he has no way to know what **safe** (15) is, nor how to do it ***accident-free*** (16)? How can he, when he is handicapped by defective materials, change of supply, machines out of order? Exhortations and platitudes are not very effective instruments of improvement in today's fierce competition, where a company must compete across national boundaries.

Something more is required. You must provide methods to help the hourly worker to improve his work, and to accomplish your exhortation toward *safe* (17) work. Meanwhile, the hourly worker sees your exhortations as cruel jokes, management unwilling to take on their responsibilities for ***safety*** (18).

A usual stumbling block in most places is management's supposition that ***safety*** (19) is something that you install, like a new Dean or a new carpet. Install it, and you have it. In your case, you handed to someone the job of Manager of ***Safety*** (20) and paid no further attention to the matter.

Another roadblock is management's supposition that the production workers are responsible for all trouble: that there would be no problems in production if only the production workers would do their jobs in the way that they know to be *safe* (21). Man's natural reaction to trouble of any kind in the production line is to blame the operators. Instead, in my experience, most problems in production have their origin in common causes, which only management can reduce or remove.

Fortunately, confusion between the two sources of trouble (common or environmental causes, and special causes) can be eliminated with almost unerring accuracy. Simple statistical charts distinguish between the two types of cause, and thus point the finger at the source and at the level of responsibility for action. These charts tell the operator when to take action to improve the *safety* (22) of his work, and when to leave it alone. Moreover, the same simple statistical tools can be used to tell management how much of the proportion of ***accidents*** (23) are chargeable to common (environmental) causes, correctable only by management.

Be it noted, though, that statistical techniques for detection of special causes alone will be ineffective and will fizzle out unless management has taken steps to improve the system. You must remove the common (environmental) causes of **accident** (24) that make it impossible for the production worker to turn out **safe** (25) work. You must remove the obstacles that separate the production worker from the possibility to take pride in his work. Failure of management to take this initial step, before teaching the production worker how to detect his own special causes, accounts, in my belief, for some of your troubles.

The benefit of this communication with the worker, if he perceives a genuine attempt on the part of management to show him what **safety** (26) is, and to hold him responsible for what he himself can govern, and not for the sins of management, is hard to overestimate.

Thus, with simple data, it is possible and usually not difficult to measure the combined effect of **accident** (27) causes on any operation.

"We rely on our experience," is the answer that came from the manager of **safety** (28) in a large company recently when I inquired how they distinguish between the two kinds of trouble (special and environmental) and on what principles. Your own people gave me the same answer.

This answer is self-incriminating – a guarantee that your company will continue to have about the same amount of trouble. There is a better way, now. Experience can be cataloged and put to use rationally only by application of statistical theory. One function of statistical methods is to design experiments and to make use of relevant experience in a way that is effective. Any claim to make use of relevant experience without a plan based on theory is a disguise for rationalization of a decision that has already been made.

In connection with special causes, I find in your company no provision to feed back to the production worker information in a form that would indicate when action on his part would be effective in helping to improve his work. Special causes can be detected only with the aid of proper statistical techniques.

Statistical aids to the production worker will require a lot of training. You must train hundreds of hourly workers in use of simple control charts.

Who will do the training? My advice is to start with competent advice and assistance for training. For expansion, search in your own ranks for people with considerable amounts of statistical knowledge and talent. Such people, taught and nurtured under competent guidance, may be able to take on training of other people. Leave that to your advisor.

There is no excuse today to hand a worker specifications that he cannot meet **safely** (29) or to put him in a position where he cannot tell whether he will be **injured** (30). Your company fails miserably here.

When a process has been brought into a state of statistical control (special causes weeded out), it has a definite capability, expressible as the economic level of *safety* (31) for that process.

There is no process, no capability, and no meaningful specifications, except in statistical control.

Fewer *injuries* (32) can be realized economically only by reduction or removal of some of the common causes of *accidents* (33) which means action on the part of management. A production worker, when he has reached statistical control, has put into the process all that he has to offer. It is up to management to provide better uniformity in incoming materials, better maintenance, change in the process, change in sequencing, or to make some other fundamental change.

In connection with the above paragraph, I find that in spite of the profusion of figures that you collect in your company, you are not discovering the main causes of *accidents* (34). Costly computers turning out volumes of records will not improve *safety* (35).

An important step, as I see it, would be for you to take a hard look at your production of figures – your so-called information system. Fewer figures and better information about your processes and their capabilities would lead to improved uniformity and greater output, all at reduced cost per unit.

I should mention also the costly fallacy held by many people in management that a consultant must know all about a process in order to work on it. All evidence is exactly the contrary. Competent men in every position, from top management to the humblest worker, know all that there is to know about their work except how to improve it. Help toward improvement can come only from outside knowledge.

Management too often supposes that they have solved their problems of *accidents* (36) by establishing a *Safety* (37) Department, and forgetting about it.

Management too often turns over to a plant manager the problems of organization for **safety** (38). Your company provides a good example. This man, dedicated to the company, wonders day to day what his job is. Is it production or *safety* (39)? He gets blamed for both. This is so because he does not understand what *safety* (40) is or how to achieve it. He is harassed day by day by problems of sanitation, pollution, health, turnover, and grievances. He is suspicious of someone from the outside, especially of a statistician talking a new language, someone not raised in the manufacturing business. He has no time for foolishness. He expects authoritative pronouncements and quick results. He finds it difficult to accustom himself to the unassuming, deliberate, scholarly approach of the statistician. The thought is horrifying to him, that he, the plant manager, is responsible for a certain amount of the trouble that plagues the plant, and that only he or someone higher up can make the necessary changes in the environment. He should, of course, undergo first of all a course of indoctrination at headquarters, with a chance to understand what *safety* (41) is and what his part in it will be.

Proper organization and competence do not necessarily increase the budget for improvement of **safety** (42) and productivity. Management is already, in most instances, paying out enough money or more for proper organization and competence, but getting tons of loss **reports** (43) full of meaningless figures – getting rooked, I'd say and blissfully at that. Your company is no exception.

Your next step will be for your top management, and all other people in management, engineering, chemistry, accounting, payroll, legal department, consumer research, to attend a four-day seminar for indoctrination in their responsibilities.

You will engage on a long-term basis a competent consultant. He will attend the seminar and guide your work on the 14 points and removal of the deadly diseases.

You should then establish appropriate organization for improvement of **safety** (44)."

REFERENCES

Original Source: "Quality Productivity & Competitive Position", MIT, Cambridge, Massachusetts, 1982

KEY TRANSITION WORDS
(WORDS FROM ORIGINAL TEXT)

1 = variable quality

2 = quality

3 = quality

4 = quality control

5 = quality control

6 = quality

7 = quality control

8 = a carload of finished product

9 = trouble

10 = out

11 = quality control

12 = quality control

13 = do perfect work

14 = doing his job better

15 = his job

16 = better

17 = perfect

18 = quality

19 = quality

20 = quality

21 = right

22 = uniformity

23 = defective material

24 = trouble

25 = good

26 = his job

27 = common

28 = quality

29 = economically

30 = he has met them

31 = quality

32 = tighter specifications

33 = trouble

34 = poor quality

35 = quality

36 = quality

37 = quality control

38 = quality

39 = quality

40 = quality

41 = quality control

42 = quality

43 = machine sheets

44 = quality

RATE YOUR 'B.O. S. S.'

LARRY L. HANSEN

Originally published as cover/feature of Professional Safety magazine June 1994

BENCHMARKING ORGANIZATIONAL SAFETY STRATEGY

Benchmarking, the planned, deliberate process of seeking and targeting competitive continuous improvement by emulating industry's 'best practices' has swept its way into today's business planning strategy. Like most progressive management techniques, competitive benchmarking was conceived in response to the challenge for improved cost effectiveness, quality and reliability in American products. Now, however it is being used to improve operating results across all business functions. Defined here are key safety benchmarks observed and documented as success drivers. Organizational issues that truly impact safety effectiveness are addressed, and a tool for 'rating' organizational issues critical to good safety outcomes is presented. Rating Benchmarks of Organizational Safety Strategy (B.O.S.S) leads to recognition that 'good' safety is 'good' management, not programs. Long-term safety improvement can only be attained by addressing organizational core competencies.

For the past 20 years, U. S. business has been victim to a national sham, one that currently drains $60 billion from national productivity. In addition, this sham has legitimatized use of legislation (the hammer) rather than education (the mine), as the predominant means of achieving workplace safety. Since Occupational Safety and Health Administration's (OSHA) inception, U. S. management has become entrenched in 'safety by compliance' rather than 'prevention by planning'. As a consequence, business now faces a sequel to the 1980's quality dilemma; American safety – doing things wrong the first time.

CURRENT PRACTICE: RELUCTANT COMPLIANCE

Today's predominant safety strategy is reluctant compliance. This translates to after-the-fact hazard detection, which (like outdated quality programs based on inspecting defects out at the end of a process) does not identify organizational errors – the true causes of accidents. The bottom line: Most safety programs are not founded in sound management theory, have negligible impact on operational efficiency, and do not contribute to corporate productivity or profitability.

These are not new or enlightening conclusions. Manager's continued resistance to 'the safety program' suggests they have recognized this reality for some time. Managers believe that safety has little impact on operations, is compliance-oriented, and impedes business outcomes. Herein lies the problem – they are correct.

As practiced today, safety does not equate with key business objectives. Line managers perceive safety as a program separate from the organizational mission, not as a 'process outcome' controlled by the management system. They fail to recognize that safety (like any other process outcome), can be improved by addressing the management structures, processes and practices that generate organizational results.

BENCHMARKING DISPELS TRADITIONAL MINDSETS

To overcome this errant perception, many progressive organizations initiate benchmarking, a process of comparative measurement, to identify and target improvement. Benchmarking challenges the myopic views held within many organizations by forcing management to look beyond current traditions and seek 'best available practices' (within and beyond their industry). This search allows a corporation to equate its own performance to that of others, thus dispelling traditional mindsets and not invented here mentalities that inhibit positive change.

The process often produces shocking realities. Corporations discover how well other organizations perform. They see the gap between 'their way' and the 'best way'. Benchmarking safety clarifies the relationship between accidents and their true sources within the management system. In addition, the process identifies organizational issues and establishes safety as a line operations function. For the first time, safety becomes management and 'is the process'.

"In successful companies, 'safe' is how things are executed continuously – not merely a topic at the monthly meeting."

LET'S RATE YOUR B.O.S.S.

Ready for a reality check? Let's remove the blinders of tradition and benchmark the safety management process. The 20 responses completing the statement: "You know your safety program is effective when..." are true drivers of safety effectiveness within an organization. Comparing current performance to these practices identifies targets for change within the management process. So, how does an organization know when its safety program is effective? Take the following test to 'Rate Your B.O. S. S'. in your organization... Good Luck!

HERE'S YOUR CHANCE TO RATE YOUR B.O.S.S.

Instructions: Assign '5 points' to each benchmark statement that accurately reflects the current 'safety reality' in your organization...all others score '0'.

YOU KNOW YOUR SAFETY PROGRAM IS EFFECTIVE WHEN:

1. _____Your Chief Executive Officer is: "CEO and Chair of the Safety Committee"

2. _____Your President says: "I'm in a safety meeting, hold all my calls!"

3. _____Your pre-tax profit exactly equals your reduction in accident costs!

4. _____Your corporation provides exotic travel incentives:

   Employees: to resort destinations...for good performance.

   Managers: to the CEO's office...for poor performance.

5. _____There are no safety costs.

6. _____Employees 'volunteer' to serve on the Safety Committee.

7. _____The 'E' in your 'E'-mission statement stands for electronic...not management's 'evacuation'.

8. _____The reduction in your Workers' Compensation cost equals the payout of your executive bonus program.

9. _____The words: Quality, Productivity, and Safety appear in the same sentence in your annual report!

10. _____Safety Committee meetings...aren't!

11. _____Employees arrive and depart from work in the same condition...smiling!

12. _____The ratio of 'Positive Recognition Notices' to 'Disciplinary Slips' in Personnel files is '8 to 1'... Positive.

13. _____Employees request a planning session on company time, at company expense, at a resort hotel… and the request is approved!

14. _____Employees have a $250 personal spending authorization for safety…no approvals required!

15. _____There is no 'time' for safety training.

16. _____You complete one full year without the word 'careless' appearing on an accident report.

17. _____Your General Manager sets a goal for 'zero accidents' and believes it!

18. _____Inspection reports don't identify 'unsafe acts and conditions'.

19. _____There is no 'safety' program.

20. _____The final phase of 'right-sizing' is complete…and you are still employed!

_____- Total Score

Scoring – Transfer your scores to the answer sheet at the end of this article to see how you've rated your organization on the 'four' drivers of organizational safety. Each area is discussed as a benchmark of excellence.

BENCHMARK # 1 – EXECUTIVE MANAGEMENT

Most executives do not know if their companies hold safety meetings…let alone attend them. A successful safety effort requires more than management commitment – it demands executive involvement. Sonoco Products Co., a Hartsville, SC packaging company reduced its corporate injury rate by 90 percent – while the national average (1975 – 1988) increased by 14 percent. How? "We learned that simply being interested in safety is not enough," explains corporate Safety Director Mike Sunderland. "Sonoco managers recognize that commitment to safety, like quality, demands more than just preaching to employees. It requires actions."

At the corporate level, Sonoco's executive safety task force is comprised of staff vice presidents and chaired by the president. Safety performance is discussed during weekly executive meetings, and any vice president whose operation has experienced an injury must present an explanatory report. "That really drives home the personal responsibility and accountability for employee safety." (Minter) At Sonoco, corporate values are established by executive action. Safety is a corporate value.

BENCHMARK # 2 - STRATEGIC VISION

Two inseparable truths are embodied in the sayings: "Safety is created in the boardroom"; and "A rotten fish stinks from the head down." Executives are being held to increasingly higher standards of care regarding personnel policies and practices. Evidence a Wall Street Journal expose implicating these values in the board-directed leadership change at Sunbeam-Oster Corp. What better way for executive management to personify a corporate safety vision than to lead the effort for employee health, safety and well-being? "Vision isn't forecasting the future, it's creating the future by taking action in the present." (Collins and Porras).

In "Safety Works, So Why Don't We Use It?" Anthony Skiff identifies how strategic direction was key to a large Ohio supermarket chain's efforts to reverse high workers' compensation loss trends. In addressing the element responsible for the turnaround, a company representative replied, "We held a series of 12 safety meetings. The President attended all 12, and had all his calls held." (Skiff). Having ultimate charge for safety within an organization is a crucial job. How many executives have personally accepted this level of responsibility?

BENCHMARK # 3 - FINANCIAL ORIENTATION

Line managers are generally uninformed about insurance costs. Perhaps 90 percent of supervisors and plant managers do not know what Experience Modification is, or their actual loss costs. This creates a problem. If managers do not understand how accidents drive insurance costs and impact operating results, they will never be motivated to address them. When training line managers in "Magic Bucks." (a term aptly describing their perception of insurance), I have found that once managers understand how losses impact 'cost of goods sold' and 'operating expenses' (two of their key responsibilities), and ultimately drain profitability, they turn on to success.

Bob Anderson, President and principal consultant of the Worksafe Group, Laguna Hills, CA confirms that every successful turnaround his company has managed with its clients was predicated by an executive awakening as to how loss costs impact operational results. Anderson recalls a large West Coast restaurant chain that wanted to reduce its $2 million annual workers' compensation losses. Upon completing their evaluation, the consultants met with the president and key operation's executives. The consultants' presentation outlined how excessive workers' compensation losses were draining corporate profits. According to Anderson, "When the president saw the financial linkage, he almost fell off his chair. He questioned the numbers four times."

When the executives finally understood the financial ramifications involved, they incorporated safety into the organization's strategic mission on quality operations and customer service. Within two years, average losses decreased 90 percent. Bottom line: Show a line manager how s/he can increase the 'bottom line' and you will find a manager who will succeed in reducing accident costs.

BENCHMARK # 4 - MEANINGFUL MEASUREMENT

DuPont Corp. has initiated numerous innovative management safety incentive and accountability programs. Occupational safety has such high priority that accident-free operations are a basic expectation of all managers. In fact, executives of operating divisions and subsidiaries must report all lost-time accidents to Dupont's CEO within hours of an occurrence.

Now the envelope has been stretched further. Sonoco Products Co. has established a corporate policy requiring plant managers to telephone the president within 60 minutes of an injury. This 'travel policy' leaves no doubt about safety's importance within these organizations. "When safety becomes a meaningful management measure, managing safety becomes meaningful." You definitely do not want to be a frequent flyer in these programs.

BENCHMARK # 5 - PARITY

Successful companies consider safety to be a long-term investment, producing significant payback through improved process efficiencies and reduced operational error. This is a crucial foundation. When managers do not believe that safety expenditures produce a positive return-on-investment, they will never provide the personal and financial support needed to elevate safety to a true operational equal.

PPG Industries, headquartered in Pittsburgh, PA recognizes the need to create safety parity (safety equal to and integrated with all key elements of the manufacturing process). The company's New Directions Program, which is based on quality management concepts addresses four core issues; 1) continuous improvement; 2) employee involvement; 3) excellence; and 4) integrated safety measures.

PPG's Crestline, OH facility achieves safety integration through a performance specification program. "At the Crestline plant, every work order now includes safety specifications, along with customer packaging, shipping, and handling specifications. The safety requirements have equal status on the work order." (Durbin). At this facility, successful job completion requires strict conformance to safe practices.

On the production floor, however, all things are not created equal. Safety can become an equal only when valued on par with, and not compromised to production demands.

BENCHMARK # 6 - EMPLOYEE INVOLVEMENT

The Cummings Jamestown Engine plant, Lakewood, NY employs a highly effective approach to employee participation and continuous improvement. The program's name 'JDIT – KAIZEN' refers to its objectives: "Just Do It - Continuous Improvement." The program incorporates several progressive concepts, including an intensive five-day employee training on continuous improvement, an emphasis on employee empowerment, and teamwork to encourage problem solving and immediate change.

Among the novel ideas incorporated in the program are 'Kaizen Carts." Mobile workstations equipped with flip charts, references and other problem-solving tools that allow employees to solve problems or evaluate process improvements on the spot. The program has produced 61 documented process improvements, leading to higher quality, lower cost, reduced cycle times, less inventory, higher efficiency, improved safety…and $250,000 in savings. "At Cummings Jamestown Engine plant, people are making the difference. They 'just do it' (Taylor and Ramsey).

This level of employee interest is not common in many companies. However, if the concept of an employee safety organization was redefined to include high visibility, adequate funding, clear responsibility, true authority, and ample recognition, many more employees would be willing to volunteer. Participation would improve 100 percent and safety programming would cost nothing. Companies would actually save money because time and efficiencies are money. Envision the potential savings when everyone thinks of a better way…and then just does it.

BENCHMARK # 7 - MISSION

Try this: Ask the next five people that pass by the safety office to read the corporate safety policy. What will they read?

1) An executive mission statement that clearly depicts a desired future (safe) state for the corporation and contains an action orientation capable of driving employee behaviors to attain it….or

2) A 'reality statement' that lacks emotion, is uninspiring, contains canned jargon, worn out phrases, and reminds employees of 'how things aren't'.

Most readers of such documents find 'number 2'. Many safety policies are boilerplate, boring and borrowed (some even still contain the originator's name), and employees know it. These policies typically contain idealistic rhetoric and conclude with: "No job is so important or schedule so rushed that we can't take the time to do it safely." Employee interpretation: "Yeah right, Why can't they design it right in the first place so I don't have to 'take time' to fix it out here?"

A successful journey begins with a specific destination in mind. If management has not created a clear road map, the journey starts now with an exercise in 'safety visioning'. If you look forward, you can see, and will create the future.

BENCHMARK # 8 - GAIN SHARING

How can meaning be added to the phrase, "Safety Pays?" If a manager's performance in controlling accident costs is to improve, build charge backs into his/her bonus and make it 'hurt so good.' Remember, what gets measured gets done, and what gets measured and rewarded, gets done well. One caution: What gets measured and rewarded gets done well, even if it is wrong. As Peter Drucker warns: "Make sure managers are doing the right things, not just doing things right." No slogans, posters, games, contests or safety incentives...please!

Why stop with executives? Why not initiate gain sharing for all employees? According to Princeton professor Alan Blinder, "Trickle Up" economics (making employee compensation contingent upon profitability) can raise labor productivity from 3 to 11 percent (Watching Out for Business). Generally, it is better to give than to receive. In this case, why not both?

BENCHMARK # 9 - SHARED VALUES

In Rome, NY, history is being rewritten...the Great Pyramids are no longer only in Egypt. The Rome Cable Corp. in Rome, NY recognizes the need for equality among quality, productivity, and safety - three critical measures of management proficiency, and emphasizes these three values in its corporate vision. Elevating safety to equal status on the production floor, and income statement must be forced through executive action.

President Shep Bayland emphasizes the values by sending clear, repetitive messages that quality; productivity and safety must co-exist for real long-term profit and sustainability to ensue. In 1992, the 'value' of safety equated to more than $212,000 (an insurance dividend earned due to reduced loss experience). At a two percent margin, that is equivalent to an additional $10 million in sales—a welcome contribution to profit and a true value.

BENCHMARK # 10 - COMPELLING MESSAGE

Safety committees are often ineffective in addressing accidents. Why? Because accidents are 'caused' by the management system, and real managers don't do safety committees. Accidents are merely symptoms of flawed operational planning rooted in the management process. Safety can only be effectively addressed in operational meetings.

Mike Gleason, Vice President of Operations for Finch Pruyn Paper Corp., Glens Falls, NY opens each operations meeting with a discussion of plant safety performance. It is no coincidence that Finch Pruyn has maintained an extremely low Experience Modification (range of .45 or lower). This affords the company significant insurance savings and a competitive advantage. In successful companies, 'safe' is how things are executed continuously…not merely the topic of a monthly meeting.

BENCHMARK # 11 - EMPLOYEE SATISFACTION

'Happy faces' are scarce in many workplaces, which signals that the safety program and most likely other aspects of the operation is in trouble. Hank Sarkis, president of the Reliability Group, a management-consulting firm, has compiled statistical data confirming 'happiness in the workplace' as the primary factor correlating with good accident experience. Hal Rosenbluth, president of Rosenbluth International Inc. and co-author of The Customer Comes Second, also identifies 'employee happiness' as the single issue from which all other organizational results flow. Rosenbluth believes that companies earn employee's poor attitudes and maintains that employee attitudes produce the organization's ultimate outcomes. "We must take employee happiness seriously. Without it, eventually everything else breaks down." (Rosenbluth and Peters).

Does this 'soft stuff' stand up to 'hard number crunching' however? Dean Witter Reynolds recently found that had investors purchased stock in companies that treated employees well, they would have earned 17 percent more than had they invested in Standard & Poor's Index Fund." (Mauer). Certainly a clear illustration of what a smile is worth!

BENCHMARK # 12 - POSITIVE RECOGNITION

The psychology of motivation has not progressed significantly in the field of management…at least not at the first-line level. Although numerous studies confirm the power of positive recognition, discipline still remains the first tool of choice for managing (a/k/a mugging) employee behavior.

Studies show that 87% of all feedback is negative. Traditional management theory has done little to rectify this situation (Kinni).

Traditional theory fails to recognize that 'discipline' is derived from the word 'disciple', hardly a negative connotation. Managers need to catch people in the act of doing something good' and tell them about it. Supervisory 'seek and destroy' missions should evolve into 'seek and reward' sessions.

In 'A Great Place to Work', Robert Levering reports how Emery Air Freight was losing $1 Million annually because employees were shipping small packages individually, rather than combining them in larger containers for less expensive bulk shipment. Management's answer: A program of positive consequences and feedback. The result: Losses were nearly eliminated. The most powerful (yet least used) motivator in the workplace…'THANKS'!

BENCHMARK # 13 - TEAMWORK

Richard Costello, manager of General Electric's (GE) Corporate Marketing Communications, challenged GE's Business Information Center to change its methods, management approaches and operational results. To help the group 'discover its future', he sent the staff (on company budget) to Epcot Center in Orlando, FL. The result: A highly effective self-directed work team. The point: When something is important to the organization, make it important to people by making people important. Is safety important to employees in your organization…and vice versa?

BENCHMARK # 14 - EMPOWERMENT = TRUST

Employee empowerment is essential to operation success. However, employees can only be empowered when funds are available to fuel innovation. Estimates suggest that 90% of all safety hazards can be corrected for less than $50, a nominal sum when compared to the average cost of an injury ($19,000+).

Zytec Corp. employees can spend up to $1000 per occasion to solve problems and/or improve operations--without executive authorization. Imagine what could be accomplished if the workforce was empowered with only 25% of that amount. Imagine employees showing supervisors how they solved safety problems, rather than waiting for crippled suggestion systems or backlogged maintenance schedules.

Zytec also employs a unique strategic management tool…Trust. And, it is not abused because the basic freedom, not the amount, is what counts. The next great reunification will involve employees and their minds. All that is needed is a little management 'seed money'.

BENCHMARK # 15 - ENABLEMENT

Although knowledge has become a key competitive strategy in industry, many still believe that safety training is a function of 'time" rather than comprehension, and that lunch periods, coffee breaks, weekends and shift changes are the only 'time' for safety. William Lareau, a noted quality author, feels that U. S. industry trains its employees to fail by basing training on duration, rather than on skills attained. Programs that provide everyone with an equal amount of training only ensure that no one's needs are fully met. In Japan, employee training differs completely…proficiency is the ultimate measure.

No one would likely debate the value of training in shaping safe work practices. Yet, managers continue to commit only minimal time to safety training. "We need a supervisory training program to deal with this 'critical issue' but it can't take more than an hour" is a common remark. For employees, the average is probably closer to 15 minutes.

Joanne McCree, human resources representative for IBM's Rochester, MN facility says: "People must be enabled as well as empowered." Unfortunately industry's track record does not mirror this advice. In a November 1992 survey of 100 small and mid-size companies, Arthur Anderson, Inc. and the National Small Business United Trade Association asked how small businesses should achieve improved productivity. The most popular answer: "Provide better training." When asked what steps these companies had actually taken to improve productivity during the past 12 months, however, training was not mentioned (If Training is the Answer…). The bottom line: Safety performance cannot genuinely change in one hour or less, as current injury statistics clearly illustrate.

BENCHMARK #16 - PROCESS IMPROVEMENT

Tom Peters identifies the phenomenon of a management awakening as "a blinding flash of the obvious." When the manufacturing process is examined, it becomes clear that management controls all vital functions, including employee selection, process design, material specification, work scheduling, environmental control and organizational culture. When accidents occur, however, the 'blinding flash' conclusion is: "Damn those careless employees." 'WRONG!'

Employees sustain injuries, accidents belong to the system and the system belongs solely to management. Audits of supervisors' accident investigation reports often reveal that more than 40 percent typically cite 'employee carelessness' as the accident cause.

To truly improve safety, start at the top, otherwise the real source of problems is being ignored. W. Edwards Deming assigns 90+ percent of all process deficiencies (including accidents) to 'common causes' – causes inherent in the system, not 'special causes' – causes attributable to individual behavior. If the company's accident investigation process is not unveiling system failure, it is not producing accurate information. The only acceptable alternative is to assign all accident causes to 'management carelessness'…and challenge them to prove otherwise. Then, the investigation system would be correct at least 90 percent of the time (which beats what industry is doing now).

BENCHMARK # 17 - CONTINUOUS IMPROVEMENT

Many companies set safety goals based on incident frequency and severity (a/k/a – Kill Rates). This process drives efforts to mediocrity -- we aspire to be average. Incident rates are nothing more than measures of how bad a company can be …and still be called good. Successful companies emphasize continuous improvement to drive results toward 'zero incidents'.

General Motors Corp., under the safety leadership of Mike Taubitz, has abandoned 'safety by the numbers' (incident rate goals), and instead focuses on continuous improvement goals and strategies. This approach eliminates tolerance of average performance, and challenges ultimate excellence. Safety, like quality, is a journey not a destination. The key: Being on the right road, heading in the right direction, and having the right speed!

BENCHMARK # 18 - PROBLEM SOURCING

More than any other activity, inspections deter organizations from attaining true safety improvement. Inspections are the core of most traditional programs, and typically consume an organization's safety focus. Unfortunately, inspections focus on the wrong sources. They address accident symptoms rather than true causes, which lie upstream in a process and are distant in both time and proximity from true causes.

Accident causes are embedded within the management system, which inspections as currently conducted, do not address. If industry would adopt Dr. Deming's philosophy and cease reliance on inspections, safety could be designed into processes. This would eliminate the efforts wasted in attempts to 'inspect out hazards', an activity that keeps many practitioner busy yet does not solve problems.

If an organization demands inspections, give them inspections…but move them off the production floor and into the executive suites, personnel office, and engineering department. With a fine tooth comb, inspect policies, directives, procedures, goals, budgets, planning processes, training programs, management development plans, staffing, organizational charts, job descriptions, communication systems, scheduling, and recognition and reward programs. Plenty of real accident causes will be found there!

BENCHMARK # 19 - ASSIMILATION (A/K/A INTEGRATION, A/K/A - COLLABORATION)

Companies with good accident experience recognize that safety is 'good business', not programs. Traditional safety programs do not exist in these companies…for one good reason—they are not needed.

The Potsdam Paper Corp. in Potsdam, NY may (on the surface) appear to lack the trappings of traditional safety programs (incentives, rules, poster, etc.). Yet the company does not experience accidents. Potsdam's approach involves participative management and teamwork—safety is a function of the process…not a separate program. As a result, employee involvement is high, hazards are controlled, and accidents don't drain productivity. All positive benefits…without a formal program.

BENCHMARK # 20 - ALIGNMENT

Congratulations! You have passed the ultimate test (unlike your more traditional peers). You have proven your ability to lead a corporation to success and savings by realizing that safety is not one thing, it is everything, and it is called ' effective management'. Welcome to the team. You'll fit in just fine…but where?

Two important points must be considered when deciding where to position safety within an organization. First, safety must be recognized as 'the management process.' Second, it is not 'where' safety responsibility is placed within an organization (whatever form it takes), but rather that it is everwhere. Progressive companies align resources to achieve results; they don't create structure. The power of influence will always outperform the power of position. If the president's voice is needed to accomplish this, serious organizational problems exist. Time to pull back and start again, this time from the bottom.

P.S. – Start packing your desk, we're moving you out of the stockroom, tool room, dispensary, guard shack or that tiny cubicle. You're one of us now…Welcome to the team!

B.O.S.S. SCORING

'FOCUSED LEADERSHIP"

1. _____ Executive Involvement

2. _____ Strategic Vision

3. _____ Financial Orientation

4. _____ Meaningful Measurement

5. _____ Parity

   _____ Subtotal 1 –5

'COMMUNICATIONS"

6. _____ Employee Participation

7. _____ Mission

8. _____ Gain-Sharing

9. _____ Shared Values

10. _____ Compelling Message

301

_____ Subtotal 6 – 10 - Communications

'HUMAN RELATIONS"

11. _____ Employee Satisfaction

12. _____ Positive Recognition

13. _____ Teamwork

14. _____ Empowerment

15. _____ Enablement

_____ Subtotal 11 – 15 – Human Relations

'INTEGRATION"

16. _____ Process Improvement

17. _____ Continuous Improvement

18. _____ Problem Sourcing

19. _____ Assimilation

20. _____ Alignment

_____ Subtotal 16 – 20 – Integration

_____ Total 1 – 20 - Your B.O.S.S. Rating

REFERENCES

Aguayo, Rafael. Dr. Deming: The American Who Taught the Japanese About Quality, New York *Fireside Publications*, 1991.

Anderson, Robert S. Personal correspondence and contributions. April 20, 1993.

"Breakthrough Performance Linking Quality Improvement to Business Results". New York, American Quality Foundation.

Burnside, James R., Letting Go, Schenectady, NY High Peaks Press, 1992.

Collins, James C. and Porras, Jerry I. "Making Impossible Dreams Come True", *Stanford Business School Magazine.* July 1989.

Competitive Benchmarking: The Path to a Leadership Position, Stamford, CT, Xerox Corp.

Durbin, Thomas J. "Safety & Quality at PPG Industries.", *Occupational Hazards.* May 1993.

"Fired Sunbeam Chief Harassed and Hazed Employees, They Say." *Wall Street Journal*, Jan. 14, 1993.

Hequet, Marc. "The Limits of Benchmarking." Training. February 1993.

Hlam, Alexander. "Closing the Quality Gap". Englewood Cliffs, NJ: Prentice Hall, 1992.

"If Training is the Answer, What's the Question?" Training. March 1993.

Kinni, Theodore B. "Motivating the Unmotivated." *Quality Digest.* March 1993.

Lareau, William. American Samurai. New Win Publishing, Inc. 1991.

Maurer, Rick. Caught in the Middle. Cambridge, MA Productivity Press, 1992.

Minter, Stephen G. "Why Sonoco is Committed to Safety." *Occupational Hazards.* January 1993.

Murray, Jay F. "Occupational Health and Safety Management: The Competitive Edge." *ENSR Insight. Vol. 1*, 1993.

Rosenbluth, Hal and Diane McFerrin Peters. "Winning Combination." *Northwest Airlines World Traveler.* February 1993.

Schulz, Randall S. and Ian C. Macmillan. "Gaining Competitive Advantage Through Human Resource Management Practices." *Human Resource Management.* Fall 1989.

Sheridan, John H. "Where Benchmarkers Go Wrong." *Industry Week.* March 15 1993.

Skiff, Anthony W. "Safety Works, So Why Don't We Use It?" *Safety Workplace.* January 1993, Vol. I Issue I.

Taylor, David L. and Ruth Karin Ramsey. "Empowering Employees to 'Just Do It." *Training and Development.* May 1993.

"Watching Out For Business." *U. S. News and World Report.* Feb. I, 1993

A UNIVERSAL MODEL FOR SAFETY EXCELLENCE

LARRY L HANSEN

For over three decades now, I've been intrigued by the 'universal question' asked by senior executives seeking to improve safety results in their organizations: "Why do employees do what they do...act unsafe and have accidents...(and file claims...and cost us so much money?)" Good question...interesting how the answer continues absent from 'B' – School curriculum!

THE UNIVERSAL QUESTION:

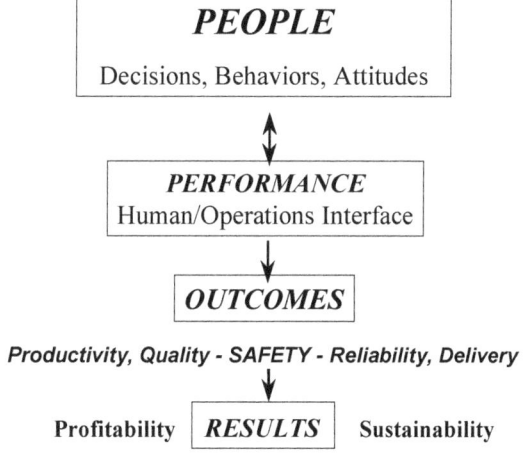

"Why do employees do what they do...act unsafe, and have accidents?"

PEOPLE
Decisions, Behaviors, Attitudes

PERFORMANCE
Human/Operations Interface

OUTCOMES

Productivity, Quality - SAFETY - Reliability, Delivery

Profitability **RESULTS** Sustainability

In 1993 being convinced that good safety programs produced but marginal results, at best, I joined the search for a better answer to this 'universal question'. Although remaining committed to 'safety', I felt it necessary to explore beyond the discipline, so I sought answers in the literature of business excellence. Premised upon D. A. Weaver's observation: " *Excellent organizations frequently achieve exceptional safety results in the absence of any visible safety program, and excellent safety performance cannot be attained in a generally poor organization."* I set out to discover the 'secret sauce' of safety excellence: "What makes great (safety) companies--great?" and "What are the defining 'elements of excellence' in safety?"

My 'Quest for (Safety) Excellence' led me in two directions at the same time...past and present. I read the history of management (past) to gain a greater understanding of how we got to where we are today, including the works of Fayol, Taylor, Mayo, Skinner and of course the contemporary views of Nikita Khrushchev embraced by most American managers today...Free*dom is good, but control is better.*

I then proceeded to read the 'contra-literature' on high performance by Dr. Deming, Juran, Conway, Peters, Covey, Senge, Stack, Collins, Welch, Drucker, and a host of others. From this pool of practical and empirical intellect, emerged eight common issues...eight differentiating 'elements of excellence' that hold the 'Universal Answer' to why employees do what they do...the 'good reasons for poor performance' in the workplace.

'THE GOOD REASONS FOR POOR PERFORMANCE'

- Unclear Vision - Destination clouded; and expectations unclear.

- Weak Values – Importance of safety questioned: deeds compromise policy.

- Poor Leadership – Absence of involvement...abdication of responsibility.

- Faulty Organization – Competing agendas, and sub-optimizing objectives.

- Poor Human Relationships – People not respected or valued...low trust.

- Inadequate Communications – Failure to share timely information and accept objective feedback.

- Inaccurate Measurement – A focus on minimizing 'numbers' rather than gaining knowledge.

- Lack of Consequences – Failure to consistently deliver meaningful consequences (+ & -) for observed behaviors.

Studying these perspectives allowed me to identify linkages, define attributes, and frame eight guidepost questions that an organization can use (most won't) to achieve peak (safety) performance. These elements have proven to be universal, as they apply across multiple business sectors (manufacturing, service, health care, transportation, profit and non-profit), and bridge the vast differences in geography, culture, and language across this planet: North America, South America, Australia, Europe, Asia, and Alabama!

Following are these universal answers in the form of a 'Universal Model of (Safety) Excellence' (with best practice citations) identifying 'What safety excellence companies 'DO' to clarify their *VISION*, establish their VALUES, demonstrate their LEADERSHIP, align their ORGANIZATIONS, build trusting RELATIONSHIPS, COMMUNICATE performance expectations, MEASURE what matters, and DELIVER CONSEQUENCES that drive desired behavior and reward desired results.

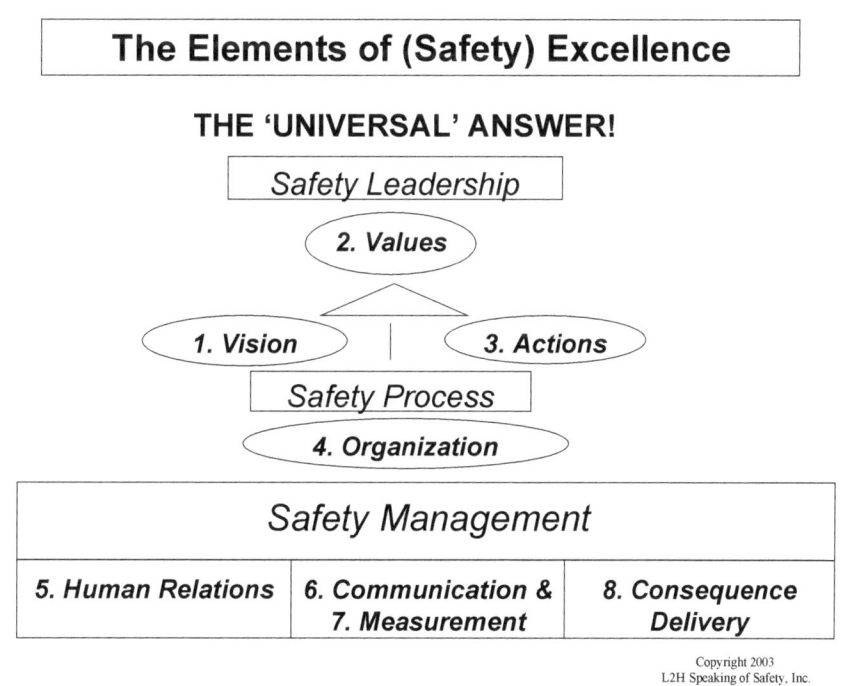

The Elements of (Safety) Excellence

THE 'UNIVERSAL' ANSWER!

Safety Leadership

2. Values

1. Vision   *3. Actions*

Safety Process

4. Organization

Safety Management

5. Human Relations	6. Communication & 7. Measurement	8. Consequence Delivery

Copyright 2003
L2H Speaking of Safety, Inc.
All Rights Reserved

Element 1. A VISION of Excellence – A clear, highly detailed, word/graphic/image, describing what a (safe) organization will 'look, sound, act, and feel ' like at a future point in time.

"I'll see it, when I believe it!"

Joel Barker, Futurist

THE POWER OF VISION

Employees perform 'safely' when the destination (vision) and road map (strategy) to (safety) excellence are clearly defined, depicted, and communicated. Excellence managers craft and communicate a vision of (safety) excellence.

- **Alcoa** – At Alcoa, 'True North' refers to its core values and vision of excellence, and in safety that means more than zero injuries. The 'True North' concept, means forward thinking - "thinking as far ahead as you can think, and then thinking further." Paul O'Neill, past CEO believed "The absence of accidents does not in any way confirm the presence of safe." According to William O'Rourke, VP of Safety, his job is to find out how the company can improve further. "We have to go past zero," he says, "We have to send employees home healthier than when they came into work." (Atkinson and Smith).

- **Calpine Corporation**. – At Calpine, leaders make their safety vision 'up close and personal' by putting senior managers in front of video cameras to capture first hand their unrehearsed core beliefs and convictions about safety for sharing with employees and the public. "We did twenty-two 30-minute interviews, and ended up with 11 hours of video which we cut down to 13 minutes," recalls Kyle B. Dotson, Vice President of Corporate Safety. "I told them it was a candid discussion to get their thoughts and feelings about safety and health in the workplace." Each executive got to say what he or she wanted to communicate. Nothing was taken out of context. The depth of feeling people had for safety surprised us (Smith 2).

QUESTION OF EXCELLENCE #1 - VISION: Have we made our expectations for, and commitment to (SAFE) operations perfectly clear to all in the organization?

Element 2. VALUES of Excellence - Those deeply held shared beliefs that define 'what's really important' in an organization'--what people are willing to go to the mat for!

"Your actions are a moving picture of your beliefs."

- Don Eckenfelder
Values Driven Safety

Employees perform 'safely' when they believe 'safety' is important to their boss, and the success of the business. Excellence managers 'Manage by Values'.

- **Levi-Strauss** – In his book 'Super Chiefs', author Robert Heller tells this story of how the power of a leader's personal 'values' drives success in an organization. At a worldwide management conference Robert Haas, Chairman of Levi-Strauss, took the podium, and read aloud an 'Aspirations Statement' (vision/values) that he believed would guide the Levi-Strauss company to success. After reading this document, he then proceeded to rip it up and toss

308

the pieces into the air acknowledging it was only paper. He then challenged all employees to seriously consider what kind of company they wanted Levi-Strauss to be, and what values they felt would make it that way. He then said, if you come up with better ones, I'll follow those. The original 'aspirations' were confirmed, and Levi-Strauss went on to be guided by them, and improved their operational and financial performance multiple times over (Heller).

- **Saint GoBain** – In order to 'manage by values', you first must have values. Bob Scherer, GM of Saint Gobain's Granville, New York Performance Materials plant has documented 19 personal values that he believes will guide his organization to 'safety success' (and they have). He personally meets with and discusses these core values with each new employee during orientation, and then continually reinforces these by conducting on going 'values affirmation' meetings with various plant departments and employee groups (Scherer).

QUESTION OF EXCELLENCE #2 - VALUES: Have we adequately identified, defined and communicated a set of core values and guiding principles that will guide us to operational (safety) excellence?

Element 3. LEADERSHIP for Excellence - The willingness and ability of key individuals and groups in an organization to make critical decisions that challenge the status quo and inspire others to follow!

"If you don't have any skin in the game, you're not in the game."

- Aussie Rules Football
Saying Thanks Kelvin Blackney

Employees perform 'safely' when leaders actively and consistently demonstrate that 'safety' is important by their decisions, deeds, and actions. Excellence mangers are 'Safety VIPs' (Visible – Involved – Participative).

- **NYS Power Authority**. – At the NYS Power Authority, President Eugene Zeltman shows up at safety meetings…all of them! Zeltman observes: "Safety requires a concerted effort by everyone, from union and nonunion workers to management…and most of all me." Zeltman attends all corporate safety committee meetings, which can last as long as two days. In fact, he has not missed one of those quarterly meetings since coming on board as CEO in 1997. "Word has filtered down to employees that if the president attends the meetings, then safety must be important," says Noel P. DesChamps, Director of Power Generation Support Services at NYPA (Atkinson and Smith).

QUESTION OF EXCELLENCE #3 – Leadership: Do our leadership decisions, actions, and inactions consistently demonstrate our values and reinforce our commitment to (SAFE) operations?

Element 4. ORGANIZATION for Excellence - The design and alignment of key roles, responsibilities, and working relationships that focus people on shared mission, collaboration, and shared rewards for achieving desired (safe) performance.

> *"Every organization is uniquely designed to exactly produce the results it achieves"*

-Peter Senge

ORGANIZATION – Employees perform 'safely' when roles, responsibilities, and relationships are clearly defined and aligned in an organization making 'SAFE' how work is done, not a program.

- **MeadWestvaco** -. Safety at MeadWestvaco starts at the top, with a Senior Leadership Team, comprised of the CEO and Chairman, President, CFO, senior vice presidents and senior-level managers, and the EH&S vice president.

  The next level is the SH&E Leadership Council - which includes corporate and facility heads of SH&E departments as well as other corporate departments, and representatives from each division of the company. This team establishes policies and provides general oversight over SH&E matters.

  Lastly, sub-network teams include safety, health and environmental professionals from throughout the company who work to implement changes within their function, and to engage in information sharing with the company's employees.

  "The results bubble back up to the SH&E Leadership Council, then go to the Leadership Team. The process allows us to align SH&E goals with company goals," says Finn Schefstad, VP Safety (Smith 1).

- **Union Pacific Railroad** - "Safety at Union Pacific is a process rather than a series of 'programs" says Steve Kenyon, General Manager-Safety. "We treat safety just as we would treat a business. Every work unit – be it a shop, an office or a track gang – must have the safety process as part of its business plan." The emphasis is on managerial as well as personal responsibility," says Dennis Duffy, executive vice president-operations, who leads one of two teams of senior managers that comprise the SHEOP (Safety, Health, Environmental and Operating Practices) Committee. The SHEOP Committee plans, directs, monitors – through visits to field sites – and adjusts the overall safety processes. The work units, however, have authority to develop and manage their own safety action plans. Determining those plans and acting on them is primarily the responsibility of the work unit's management, working with safety captains and safety committees. Both are crucial components of UP's overall safety structure (Smith 2).

QUESTION OF EXCELLENCE #4 – Organization: *Have we designed and aligned roles, responsibilities, and relationships so that collaboration toward common goals is integral to our processes, practices, and rewards?*

Element 5. HUMAN RELATIONSHIPS for Excellence: Policies, procedures, and practices that respect and place a high value on people, strengthen the bonds between employees and their company, and build TRUST.

> ### *"People just want their rights until you trample all over them...then they want revenge."*

> - Phillip Crosby
> "Quality is Free"

- **Johnson & Johnson** – At J&J, safety is embodied in its CREDO—its guiding principles of business success. Sixty years ago, Johnson & Johnson published a statement declaring working conditions must be "clean, orderly and safe." The company continues to stand by that credo for 108,300 employees at its facilities in the United States and in 54 countries around the world. The second paragraph of the credo is very specific with respect to employees. It states, "We are responsible to our employees, the men and women who work with us throughout the world. Everyone must be considered as an individual. We must respect their dignity and recognize their merit. They must have a sense of security in their jobs. Compensation must be fair and adequate, *and working conditions clean, orderly and safe.* "Safety and health are our highest values," says Joseph Van Houten, Ph.D., CSP, Worldwide Director of Planning, Process Design and Delivery, Johnson & Johnson Safety & Industrial Hygiene. "The ambitious objectives of profits, sales and production do not in any way diminish the importance the company places on safety as a measure of excellence" (Smith 1).

- **Dow Chemical -** In my safety excellence facilitations I frequently ask two important questions: 1. What is the difference between managing and leading, and 2. What is the common denominator of both? The common denominator is 'POWER'. Managers have power of position, (granted from above) while leaders have power of influence, (earned from below), and what often defines the difference in performance between excellence companies and all the rest, is how it is used.
  Managers hold power, (Manglers (not a typo) abuse power), and leaders give power freely... they empower others. In his presentation, 'Empowering High Performance at Dow Chemical', at the 2004 ASSE Symposium on World-Class Safety, Donald S. Jones Sr., VP Safety presented these convincing results. Using Total Incident Rate (TIR) as a metric, those Dow facilities that had not yet embraced empowerment (manager directed) generated incident rates

311

averaging: 4.47. Those plants that had achieved Stage 1 status (partial implementation) had average rates of 1.16. And those facilities that had accomplished Stage 2 (full implementation) had an averaged incident rate of .62. Excellence companies perceive people as the solution… not the problem. Power to the People…and profits to shareholders (Jones).

Element 6. COMMUNICATIONS for Excellence: Messaging systems (and practices) that assure timely dissemination of information, and unfiltered feedback systems that allow for the discovery of hidden truths within an organization.

> *"Employees are in the best position to prevent loss, but they need open channels to share their ideas."*

-Tillinghast – Towers Perrin WC Study

COMMUNICATIONS – Employees perform 'safely' when communication systems and practices clearly establish expectations, provide timely information, allow undistorted feedback.

- **Sempra Energy Utilities** – The Sempra Energy safety staff, guided by Jim Noon, Diane Kenney (D1), and Diane Mann (D2) and Sempra's senior leadership team host annual Safety Congresses for both their SDG&E and SoCal Gas managers, safety support staff, and safety committee members (champions). The full day agenda of each Congress consists of 'best practice' technical sessions developed by employees, and open dialogue sessions with members of the senior management team. Sempra recognizes that direct, undistorted 'face to face', and accountable communication not only looks good…but also works best.

- **Steelcase Corp** – of Zeeland MI, sends a very clear message that safety is not a 'competitive advantage' for their organization…it is rather a 'collaborative advantage'. Steelcase believes that health and safety is so important to their organization that they dedicate a full day communicating that importance each year at a safety conference for 'all managers'. In 2004, recognizing an opportunity to further 'spread the word', Steelcase expanded its invitation to include their vendors, service providers, members of the local business and professional safety community, and their competitors--but 'no cameras' please.

QUESTION OF EXCELLENCE #6 - Communications: Have we created adequate messaging and information systems (and practices) that facilitate (SAFE) decision-making and support effective problem solving in our organization?

Element 7. MEASUREMENT for Excellence: The key metrics that communicate: 1. What's really important in an organization, 2. Whether you're winning or losing the game, and 3. Ultimately, if you'll 'stay in the game'!

"Numbers are numbers...numbers are not knowledge."

- W. Edwards Deming.

MEASUREMENT - Employees perform 'safely' when the metrics upon which they are measured make 'safety' important to the business, and a key factor in their performance rating.

- **Proctor & Gamble** – P&G, under past leadership of Rick Fulweiler, transformed the EH&S function from a cost center to a 'profit center' by defining safety's contribution to the 'bottom line'. The EH&S function identifies industry average incident rates for their operations using BLS data, and tracks the average cost of lost time and recordable incidents using industry cost data. They then calculate (# of accidents prevented X average cost/incident saved), and identify how much was saved (compared to average) and thereby dropped to the bottom line as 'margin contribution'. P&G knows that 'Safety Pays'.

- **Active Agenda, Inc**. - Dan Zahlis (President of Active Agenda, Inc.) in a past life as the Western Region Risk Manager for the Häagen-Daz Company, developed what he refers to as *'the ultimate metric of safety'*.

  Instead of measuring 'recordable rates' (which created an atmosphere of fear and underreporting), he measured, and rewarded 'Total Cost per Incident' – calculated as: "Total Injury Costs - divided by - the Total of ALL Incidents (near Hits, first aid, medical only, and lost time & restricted). By design of this metric, the only way the operation could truly improve was to either drive down injury costs (by managing people better), or drive up incidents reported (to learn more about risks). Encouraging 'truth' removed the 'veil of fear', and dramatically reduced the Division's total Injury costs. (Zahlis)

Note: - Active Agenda is an open source 'FREE' Risk Management technology project. Visit www. ActiveAgenda.net to learn more about this powerful Risk Management Operating System.

QUESTION OF EXCELLENCE #7 - Are we 'gaining knowledge' by measuring what matters and valuing what counts? Or are we 'fueling ignorance' by incentivising 'rates' that defer reporting, distort truth, and impede prevention…(until it goes boom!).

Element 8. CONSEQUENCES for Excellence: Performance management systems and practices, which effectively recognize, respond, and reinforce desired (SAFE) decisions, actions, and behaviors…at all levels of an organization.

"You simply cannot manage yourself out of problems you behave yourself into."

-

Stephen Covey
Seven Habits

CONSEQUENCE DELIVERY – Employees perform 'safely' when significant and meaningful consequences (positive & negative) attach to their (safety) performance and rewards.

- **Bronson Healthcare Group** in Kalamazoo, Mich., asks all managers to write twelve thank-you notes per quarter, and to show them to their own managers as proof that they were indeed recognizing their employees. Additionally, human resources does random spot-checks on managers, asking to see copies of thank-you notes, and if a manager doesn't have them, he or she is asked to schedule a "little talk" with the senior leader of the group. They've never had to schedule more than one talk before managers quickly got the message that the organization was serious about this activity (Nelson).

- **Haynes International, Inc.** – President Francis Petro believes in commitment with accountability. "Every single manager is evaluated first on safety before anything else, and that is written and documented." "When there's an accident, a spill, or an environmental release, managers must call my office immediately and provide written documentation within 24 hours."

  Petro also reinforces that accountability through discipline. He has placed managers on probation or given them time off without pay and disciplined foreman and supervisors for not following safety procedures. "If you ever, ever, ever put production before safety here, you won't be employed here," says Petro. "We tell people safety is not a policy. That you have to behave it, think it, breathe it and act safely all the time." "It has to be in every aspect of what you do. You have to be tenacious. Every place I go, everything I do, I start the conversation with safety" (Smith 1).

QUESTION OF EXCELLENCE #8 – Consequence Delivery: Do we systemically recognize, reinforce, and reward (SAFE) behavior equal to all other parameters of operational performance?

To achieve peak (safety) performance, we must manage beyond programs that evade 'the good reasons for poor performance', and address the systemic and synergistic 'elements of excellence' that determine what's really important, and ultimately how work gets done ...safe vs. unsafe in an organization. We must manage the:

SYSTEM FOR EXCELLENCE:

- A clear (safety) **VISION** requisite to success.

- Safety as a core **VALUE** supported by guiding principles.

- **LEADER ACTIONS** that impact followers.

- **ORGANIZATION** of roles and responsibilities to assure collaboration

- **HUMAN RELATIONSHIPS** that build trust.

- **COMMUNICATIONS** that inform and enable.

- **MEASUREMENT** that encourages and tracks the right things.

- Delivery of **CONSEQUENCES** that effectively manage performance.

The Universal Model of Safety Excellence

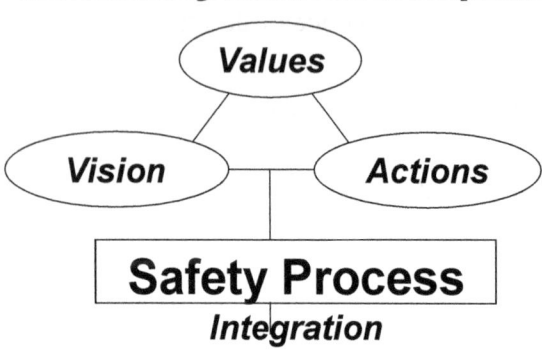

Safety Leadership

Values

Vision *Actions*

Safety Process
Integration

Safety Management

Human Relations	Communication & Measurement	Consequences

PEOPLE
Beliefs, Attitudes, and Decisions

BEHAVIORS
(Actions/Inactions)

PERFORMANCE
Human/Organizational Interface

OUTCOMES
Quantity, Quality, Efficiency, SAFETY, Productivity, Delivery, Creativity

RESULTS
Profitability & Sustainability!

REFERENCES

Atkinson, William, and Smith, Sandy, America's Safest Companies – 2002, *Occupational Hazards Magazine*, Penton Publishing, Sept. 2002.

Deming, W. Edwards, "Quality, Productivity, and Competitive Position", Cambridge, MA, MIT Center for Advanced Engineering, 1982.

Drucker, Peter, F. "Technology, Management, and Society", New York: Harper and Row Publishers, 1972.

Eckenfelder, Donald J. "Values Driven Safety", Government Institutes, Inc., Rockville, MD, 1996.

Hansen, Larry L., "The Architecture of Safety Excellence", *Professional Safety*, May 2000.

Heller, Robert, "Super Chiefs", Truman Talley Books/Dutton, New York, 1992

Jones, Empowering High Performance at Dow Chemical Presentation at American Society of Safety Engineers, *Symposium of World-Class Safety*, New Orleans, March 2004.

Nelson, Bob (The Guru of Thank You) Website Resources and posted articles (www.nelson-motivation.com).

Scherer, Bob, Saint GoBain Performance Materials, Presentation at Business Excellence Conference October 2005.

Sempra Energy Utilities, SoCal Gas 2005 Safety Congress, Compton, CA.

Smith (1), Sandy, America's Safest Companies – 2003, *Occupational Hazards Magazine*, Penton Publishing, Sept. 2003.

Smith (2), Sandy, America's Safest Companies Share a Passion for Safety – 2004, *Occupational Hazards Magazine*, Penton Publishing, October 26, 2005.

Steelcase Corporate Manager's Conference, 2004 Zeeland MI.

Tillinghast-Towers Perrin Survey Report, "Assessing the Performance of Workers' compensaiton Cost" - *Management Initiatives* 1996.

Zahlis, Daniel F, and Hansen, Larry L., "Beware the DISCONNECT", *Professional Safety,* November, 2005.

A UNIVERSAL MODEL FOR SAFETY X-CELLENCE.

SECTION 14

Technology – Active Agenda

*"Priority is better achieved through the
ability to capitalize on similarities
than the ability to highlight differences."*

~ Dan Zahlis, Author
"The Hidden Agenda"

"Info, info, everywhere, but no one stops to think."

— H. Krantzberg

THE SOLUTION REVOLUTION

DANIEL F. ZAHLIS, LARRY L. HANSEN, AND MATTIAS J. THORSLUND

Originally published in the October, 2006 edition of Occupational Hazards magazine

"There are risks and costs to a program of action. But they are far less than the long-range risks and costs of comfortable inaction."

- John F. Kennedy

Following more than 50 years of combined experience in risk management, and 10 years of searching for a shared solution to the global loss problem, we have reached this disconcerting conclusion: multiple forms of individual, organizational and market imposed control of data and resources, driven by proprietary motives are impeding success in the safety and risk management profession.

To overcome market-imposed controls, risk management professionals will need unrestricted access to the tools, collaboration and support of a global network of human and technical resources. We propose a global solutions project - the Active Agenda - that will enable inter-functional and inter-disciplinary collaboration, reinforce corporate values and optimize organizational efficiency and results among community members. We propose "a Solution (R)evolution," which would make the product of this collaboration available to all - for free!

Following a 10-year quest to unearth a solution capable of resolving the problem - the pain of human injury and billions of dollars pouring out of organizations every year - we've come a conclusion: To succeed in our profession, we must set our companies, our customers, our colleagues and our competitors free!

Speaking to Fast Company magazine on the courage requisite for effective leadership, Dr. Robert Jarvik, inventor of the Jarvik-7 artificial heart said," *Leaders are visionaries with a poorly developed sense of fear, and no concept of the odds against them. They make the impossible happen.*"

We don't claim to be visionaries. However, we fearlessly are committed to succeed in our mission to inspire a global project, based on an open and mutual network that will enable the safety and risk Management profession to succeed - against all odds.

This article is about the challenges and systemic impediments encountered in bringing innovative change to the safety and risk management professions that are controlled by market interests that can only thrive if the profession fails.

It also is an open invitation to operations, risk and safety managers to participate in a freedom march which will break the chains of control that perpetuate both human suffering and economic loss in the workplace.

If you are a person of conscience concerned with the escalating cost of loss and frustrated by the lack of progress in stemming this huge drain on people and profits, then read on: This message is for you!

THE (EVOLVING) PROBLEM

It is often said that 90+ percent of problem solving lies in accurate problem 'seeing.' We have found this to be the case in operational risk control.

Early on, the profession perceived the problem to be: "How do we achieve statutory compliance and avoid regulatory fines?" As time progressed, the problem was redefined to: "How do we safeguard facilities to limit exposure?" It then evolved to: "How do we change poor employee attitudes," and then it evolved into a process of observing, measuring and managing at-risk behaviors.

As the profession's understanding of business motives improved, it morphed further, to: "How do we manage lagging performance indicators to impact 'the numbers'?" Now, the problem has become: "How do we stop the bleeding - the escalating costs of injuries?"

Efforts to address each incarnation of 'the problem' have been met with a common approach - imposed controls, some external and some internal. These controls were based on command authority: rules, policy, compliance hammers, corporate mandates, progressive discipline and, of course, the obligatory training, re-training, remedial training and disciplinary training, all of which disregarded the real sources and causes of workplace accidents: dysfunctional management systems!

After years of best efforts, and with senior managers growing impatient for results 'now,' the core impediment has become clear and ironically, it is a basic tenet of the safety profession - control-oriented thinking.

Whether motivated internally or externally, control strategies ultimately impede performance by shutting down communications, eroding trust, eliminating cooperation and sequentially transforming clients into 'hosts,' internal functions into 'competitors,' and workers into 'adversaries.' Imposed controls also promote the transformation of employees from valued coworkers to victims, then patients, then claimants and eventually, to litigants or 'enemies.'

321

We have been so focused on problem solving (via control strategies), that we have failed at problem seeing. Our attempts to control numbers have reduced our access to preemptive organizational knowledge (culture indicators) and have severely suppressed availability of organizational data (low-severity incidents). Our preoccupation with control as a solution has prevented us from seeing freedom (a.k.a. transparent collaboration) as the solution.

THE KEYS TO SUCCESS

The keys to success lie in those functions favored within organizations, such as operations, manufacturing and engineering. In viewing these, it becomes obvious that their similarities are far more plentiful than are their differences, with the problems of waste, expense and loss costs being the common enemy across functions.

Risk management success lies in our ability to align with these core functions, and leverage their organizational advantage (power, position and resources) to achieve shared business objectives. This is best accomplished with collaborative systems that eliminate process waste (redundancy and inefficiency), reduce loss costs (claims and expenses) and promote trusting relationships based on values that strengthen organizational cultures and lead to greater productivity and sustainability.

Recognizing that in business, numbers define performance, and that wrong numbers can defeat best efforts, an integrated risk management system must utilize integrated metrics that are sensitive to functional differences, yet focused on common financial objectives.

At a certified risk managers course presented by the Insurance Institute of America, we were intrigued by a session that proposed that "insurance was only 20 percent of 20 percent of risk management." Voila! Finally, it made sense! The 'tail' (financing) was wagging the 'dog' (risk identification, measurement, control and administration - a.k.a., operations). No wonder the 'dog' was confused, frustrated and starved for results.

The objective is now clear. As a profession, we must leverage the power of technology to create an integrated risk management operating system capable of enhancing collaboration, reinforcing values and optimizing organizational efficiency and results. We must make this technology available to all for free, and invite everyone to share and improve the evolving solution!

One might ask: If open, transparent management and collaborative knowledge sharing via today's technologies is the solution, why then isn't such a solution available today?

And that answer is simple. Efforts to inspire collaboration and freely distribute innovation in the safety and risk management profession run counter to proprietary (control-oriented) motives imposed by institutions that benefit from problems (loss) remaining unsolved. These motives are many, each driven by self-serving, proprietary interests, and each the enemy of the Active Agenda.

HARSH REALITIES

There are several harsh realities and hidden agendas created by, and embedded in the risk and safety marketplace and opportunities to expose and confront them available through the Active Agenda project.

Harsh Reality #1: Insurance carriers retain ownership of client loss information (which aids in control of the client).

The Active Agenda is open and not controlled by any individual or institution. The Active Agenda retains ownership of knowledge and experience (a.k.a. data) within the organization where it is generated.

Harsh Reality #2: Insurance brokerages are resistant to value-adding innovations that may threaten relationship marketing (a.k.a., commission control).

The Active Agenda requires risk services firms such as brokers and consultants to differentiate themselves based on the value of their services, in support of the client organization's best interests. The Active Agenda demands value from service providers.

Harsh Reality #3: Insurance brokers and carriers are motivated to control clients via renewals and retentions by using proprietary products, practices and services.

The Active Agenda changes relationships from dependency on proprietary products and services to procurement of value-adding support services and risk financing vehicles. The Active Agenda aligns the interests of the enterprise with the reality of the insurance industry business model - client-controlled solutions vs. carrier investment income.

Harsh Reality #4: Self-insured companies rely on the same or similar proprietary services offered to traditionally insured companies through insurance brokers and service providers. Self-insured companies are susceptible to similar forms of control.

The Active Agenda puts the risk managers of all functions, at all levels, in control of company data, and postures them to identify and manage risk, negotiate better programs and maximize the impact of risk control investments.

Harsh Reality #5: OSHA's record-keeping requirements and targeting based on rates practices

323

unintentionally misguide organizations by encouraging the suppression of data that is required for good decisions and improved results. The end result is lower numbers, and higher loss cost.

The Active Agenda enables the measurement and analysis of pre-incident activities performed to preempt incidents and reduce costs. The Active Agenda utilizes tools to capture high frequency reporting of low-severity events, and inspire measurable actions.

Harsh Reality #6: IT departments can be impediments to improved performance by their support decisions. These decisions can lead to control of innovation.

The Active Agenda utilizes open source code that can be downloaded, customized, maintained, hosted, improved and supported by local users and IT departments. The Active Agenda eliminates the need to ask permission for success from a vendor or service provider of any kind.

Harsh Reality #7: High turnover among risk control service providers results in business discontinuity, and a series of demoralizing restarts.

The Active Agenda plans for this by capturing upstream and downstream risk control participation and activity. This ever-growing database of retrievable knowledge and history can substantially ease the impact of internal staff changes and the revolving door of external risk control service providers (often resulting from internal staff change - see the cycle?).

Harsh Reality #8: Many companies profess: "Safety is our No. 1 priority," but, in reality, safety initiatives are frequently under-funded due to operational budgetary constraints and the perception that safety is not part of the core business.

The Active Agenda is free and open! It is comprised of tools that users are free to download, free to use, free to share, free to improve and free to integrate into core business tasks. The need to compete for scarce resources and seek budget approvals for proprietary products is eliminated.

ENABLING CORE TRUTHS

Here are some of the core truths of high-performing organizations and characteristics of the Active Agenda required to enable the conversion of core truths to practical realities:

CORE TRUTH #1: Risk control is integrated into core business processes.

The Active Agenda manages business objectives utilizing shared processes. The Active Agenda flattens operational risk management by collapsing functional, organizational and geographical silos

324

and streamlining risk control with shared resources and common business processes.

Core Truth #2: Ownership of process is key to success.

The Active Agenda tracks internal and external accountabilities and involvement. The Active Agenda tracks, measures and reviews service provider promises, involvement, delivery and performance accountability.

CORE TRUTH #3: Collaborative business processes yield better results.

The Active Agenda collects data from across the global enterprise and allows open access to data thereby allowing organizations to more effectively collaborate and impact performance.

The Active Agenda enables resource availability and deployment by providing reciprocal access between the organization and a globally dispersed network of best-qualified solution providers.

The Active Agenda allows advisory support from an unlimited number of invited subject matter experts and incorporates the insightful guidance of experienced people and organizations - no matter where they are or which organization they report to.

CORE TRUTH #4: Problems AND solutions scale.

The Active Agenda deploys solutions that are equally effective for a single department, of a single physical location, of a single organization and for multiple departments, of multiple physical locations, of a global network of organizations.

CORE TRUTH #5: Process measures upstream are better metrics for controlling loss downstream.

The Active Agenda captures and benchmarks performance data, upstream of loss. The Active Agenda focuses on process-related metrics and calibrates against results.

CORE TRUTH #6: Open access to organizational knowledge yields better results than hidden truths.

The Active Agenda deploys solutions with complete transparency, sharing risk-control methods and loss experiences with a global network of workplace protection stakeholders. Open access allows decentralized resources, stakeholders and service providers to observe and assist with risk control

325

from wherever an Internet connection is available.

CORE TRUTH #7: Business culture drives loss costs…and just about everything else in an organization. Collaborative processes increase communication, improve performance and reduce waste.

The Active Agenda enables the collaborative management of business risk by centralizing business tasks and processes and decentralizing solution providers. The Active Agenda relies on collaborative processes to enhance culture and reduce loss costs.

The Active Agenda seeks the tools necessary to consider migration to the alternative insurance market where the organization is in a better position to manage its interpersonal, interoperational, interorganizational and interglobal relationships. The alternative market helps organizations foster a more positive, proactive culture.

The Active Agenda targets business practices and cultural improvements that are not found in the traditional risk control agenda. The Active Agenda increases the utilization of public forums, surveys, feedback and the tracking of threats to business values.

A SOLUTION (R)EVOLUTION

For risk control professionals to succeed in driving change in their organizations, they will need the tools, support and collaboration of an open network of human and technical resources committed to confronting those forces positioned to impede their progress.

The solution is freedom; freedom to openly use, share, distribute, modify and benefit from the solutions of a global network of organizations and individuals that are more interested in solving the problem than profiting from it.

That solution is Active Agenda - an open source, 80-module (for now), Web-enabled, risk management operating system and global collaboration project available free under the reciprocal public license.

Our efforts to persuade an industry from the top down have failed. If we are to succeed in our vision, mission and commitment, we will need to invoke and inspire a Solution (R)evolution.

Will you join us?

Visit http://www.activeagenda.net to enlist in a movement that can change the future of risk and safety management.

Risk and Insurance Management Society. *1998 RIMS Benchmark Survey.* New York: RIMS, 1998.

Zahlis, Dan F. "Caution: Beware OSHA Statistics", Professional Safety, December 1995.

DAN ZAHLIS

OPEN ENTERPRISE RISK CONTROL: THE ACTIVE AGENDA PROJECT

INTRODUCTION

Active Agenda is a browser-based, cross enterprise, open source, risk management software application capable of reducing the billions of reported dollars lost each year due to workplace injuries and the billions of unreported dollars spent to administrate failed loss control programs.

The application utilizes Web technologies to link and enable operational managers, risk control practitioners, and external vendors to collaborate for improved risk control results. With Active Agenda, companies can identify and **substantially reduce the Total Cost of Risk**[1], including risks to employees, visitors, customers, property, the environment, products, vehicles, employment practices, security, and other operational exposures.

BACKGROUND

Active Agenda was conceptualized in 1995 while its founder, Daniel F. Zahlis, was working as the Western Region Risk Manager for The Häagen-Dazs Company. Zahlis was employed to develop and formalized the company's risk control structure and process.

---

[1] 1998 RIMS Benchmark Survey. In 1993 the cost of risk concept was formalized by the Institute of Management Accountants and RIMS with the publication of a standard of management accounting statement titled Practices and Techniques: Internal Accounting and Classification of Risk Management Costs.

The goal of the statement was to provide risk managers with a consistent and comparable method of accounting for risks. The statement defines risk management costs as applying to three exposures: property, tort liability, and occupational disease or injury. Within these three exposures are six types of costs: insurance premiums; retained losses; internal administration; outside services; financial guarantees; and fees, taxes, and other similar expenses.

After a period of identification, measurement, and analysis, Zahlis concluded that Häagen-Dazs' distributed responsibilities for the employee safety program were more effective than more traditional (centralized) programs. With his experience leading 'resource lean" operations, Zahlis formalized the existing structure of distributed ownership and leveraged many of the effective control processes already being implemented by other organizational functions. Zahlis also evaluated the legal mandates and corporate requirements assigned to operating locations by the corporate headquarters. His objective was to identify redundant business rules, and build interdependent processes that would satisfy each mandate, and corporate requirement. This resulted in integrated business processes that eliminated wasteful redundancy across all functions of the organization.

Integrating business practices proved to be highly effective, but the organization lacked the ability to capture and distribute risk control knowledge across the enterprise. This deficiency limited the organization's ability to extend local successes across the global enterprise.

RATIONALE

In 1995, Zahlis began looking for an application capable of capturing, centralizing, measuring, and sharing operational risk information across a global enterprise. The concept for Active Agenda was born when his search revealed only fragmented pieces of a solution.

Zahlis discovered that risk control software firms had a narrow focus, and did not address the interdependent sources of risk across an enterprise. These companies offered portions of a solution in a disjointed or siloed manner. These solutions did not address the inherent redundancies and discontinuities that waste organizational resources and increase the Total Cost of Risk. These proprietary solutions were incapable of enabling broad collaboration, and were difficult to customize, modify, or tailor to the specific needs of an organization.

ALIGNING DESIGN, DEVELOPMENT, AND DISTRIBUTION

While searching for supporters for the project, Zahlis was introduced to Larry L. Hansen. Hansen had spent thirty-three years working for Wausau Insurance Companies.

As Zahlis and Hansen discussed and concurred on industry impediments to risk control innovation, Active Agenda, Inc.'s Product Architect, Mattias J. Thorslund suggested Open Source distribution as a method to circumvent the many barriers to innovation and distribution. The three began exploring Open Source software development, and immediately recognized the importance of Open Source philosophies, and their ability to impact thinking in the risk and safety profession. In 2003, they began to forge Active Agenda as an Open Source software development project.

Today's Active Agenda was designed to address the following Core Truths of High Performance and Risk and Safety Market Realities by incorporating Open Source development and distribution principles.

CORE TRUTHS OF HIGH PERFORMANCE:

⇨   Safety is integrated into core business processes.

- Active Agenda is designed to manage business objectives utilizing shared processes. Active Agenda 'flattens' operational risk management by collapsing functional risk silos and streamlining risk control.

- Active Agenda never mentions 'safety' by name. It consists of 50+ modules to manage risk by method - not function.

- Active Agenda utilizes the assignment of one or more 'risk imperatives' with every record of every module. This method allows organizations to manage cross-functional risks using single, centralized systems and generates a single view of risk control performance by risk imperative. The number of risk imperatives is unlimited and typically include functional business risks; such as People Safety, Product Quality, Operations, Labor Relations, Environment, Security, Ethics, and Financial Protection.

⇨   Ownership is key to success.

- Active Agenda allows organizations to track accountabilities and involvement. Accountable persons and associated accountabilities can be displayed within a centralized view that can be filtered, charted, measured and reviewed. In fact, accountabilities can be reassigned on a person-by-person, accountability-by-accountability basis, or as a mass reassignment between outgoing and incoming risk control human resources.

⇨   Collaborative business processes yield better results.

- Active Agenda captures data from across the global enterprise or participation base, allowing organizations to collaborate via open access to organizational issues that have an impact on local performance.

- Active Agenda's Partnerships module encourages process sharing by centralizing expectations and posting compliance methods for all to benefit from.

330

- Active Agenda utilizes a single, user defined, risk matrix that is used across the application to estimate risk likelihood and severity. Record likelihood and severity result in a calculated value that automatically returns an organizational Risk Recommendation based on a simple multiplication table. The same risk matrix is used for risks of all types.

⇨ Problems AND solutions scale.

- Active Agenda can be used by a single department of a single physical location; OR,

- Active Agenda installations can include multiple locations of a single organization; OR,

- Active Agenda can include multiple, unrelated organizations based on homogeneous risks.

- Active Agenda can start in a single department, of a single physical location, and extend to include additional participants as comfort levels and successes increase.

⇨ Process measures (upstream) are better for controlling loss than downstream metrics.

- Active Agenda enables benchmarking of performance data, upstream of loss (and subsequent to loss), based on facility, geographical location, and/or industry.

- Every module of Active Agenda can be used to generate process related metrics (e.g. Average Loss Cost per Hazard Reported)

⇨ Open access to organizational knowledge yields better results.

- Active Agenda allows organizations to deploy with complete transparency, sharing 100% of risk control methods and loss experience. However, it is also possible to limit each user's and/or organization's access to data.

- Active Agenda captures preemptive risk control activity data and makes it available to anyone possessing user credentials and access to the application. This method allows centralized resources, stakeholders, and service providers to observe and assist with risk control from wherever an Internet connection is available.

⇨ Business culture drives loss costs.

- Active Agenda enables the collaborative management of business risk by centralizing business tasks and processes designed to identify, measure, analyze and report risk. Collaborative processes increase communication, improve performance and reduce waste. Collaborative processes enhance culture and reduce loss costs.

- Active Agenda's Loss Costs module allows organizations to track costs by specific categories that allow the organization to manage the 'claims manager.' Claims managers and/or adjusters can represent a significant threat to business culture. Claims management practices are reflected in claims costs but most adjusting firms don't categorize costs at a level that empowers an insured to identify costly practices. Active Agenda can track detailed costs categories and trend adjuster practices that increase costs.

- Active Agenda provides the tools necessary for companies to consider migration to the alternative market where they are in a better position to manage relationships between the organization and those filing loss claims against it. The alternative market helps organizations foster a positive cultural climate.

⇨ Certain loss characteristics reflect and present risk to the business culture. Active Agenda's Loss Characteristics module allows organizations to analyze loss claims, perform trend analysis by loss characteristics, and set triggers for characteristics that are predictive of culturally driven loss costs (e.g. litigation).

- Active Agenda's Outside Counsel module tracks litigation and allows organizations to assess cost trends associated with law firms – both plaintiff (applicant) and defense. Litigation trends can reflect a business culture at risk and litigation substantially increases cost.

- Active Agenda possesses many modules that target business culture improvement as a means of reducing loss costs. These modules address topics and business processes that are not found in traditional risk control software products. These modules include, but are not limited to: Incentives, Suggestions, Town Hall (public forum), Surveys, Feedback, and Values Threats (tracking threats to corporate values).

RISK AND SAFETY MARKET REALITIES:

⇨ Insurance carriers spend less than 1½% of collected premium on worker safety initiatives.

- Active Agenda is Open Source and not controlled by any institution. Based on the obvious lack of financial commitment by insurance carriers to prevent loss, Active Agenda would not be built by a carrier.

332

- Active Agenda returns ownership of data (and databases) to the organizations generating the data. This allows organizations to easily modify risk financing methods based on price and value added services of the institution providing the financing vehicle.

⇨ Insurance brokerage firms are highly resistant to any innovation that may threaten their commission base.

- Active Agenda allows organizations to capture data capable of generating insurance or reinsurance carrier quotes without the need for an insurance broker; however,

- Active Agenda allows risk services firms (i.e. brokers and consultants) to differentiate themselves based on the value of their client services in support of Active Agenda and their clients' best interests.

⇨ Companies that purchase insurance rely heavily upon the experience, recommendations, and services offered by insurance brokers and insurance carrier loss control consultants. These service providers are motivated to control clients using proprietary products and services.

- Active Agenda shifts the insurance carrier differentiation model from proprietary products and services, to support services and financial vehicles. Active Agenda aligns the purpose for insurance with the reality of carrier investments.

- Active Agenda increases the availability of support to deploying organizations through discussion forums, free support programs, distributed and independent service providers, and other global community support mechanisms.

⇨ Self-insured companies, and other businesses financing their risk through the alternative market, rely on the same or similar services offered to traditionally insured companies through their insurance brokers.

- Active Agenda puts the Risk Manager in control of company data and postures them to negotiate better premiums based on more reliable, upstream data.

- Active Agenda allows Risk Managers to maximize the impact of risk control investments by targeting risk control services to support organizationally owned systems and processes.

⇨ IT departments can be impediments to innovation.

- Active Agenda's source code is open and can be downloaded, customized, maintained, and/or supported by local IT departments. Active Agenda's success as a product is not dependent on Active Agenda Inc.'s success as a company.

- Active Agenda is browser-based and can be hosted any way an IT department is most comfortable. Active Agenda can be run from a single installation, on a single server (computer), sitting next to a single person, serving a single purpose; OR,

- Active Agenda can be hosted on a local Intranet or in a hosted environment to circumvent the obstruction of an IT department.

⇨ High turnover rates among risk control and service provider employees result in business discontinuity.

- Active Agenda captures an organization's upstream and downstream risk control activities. This ever-growing database of risk control knowledge can substantially ease the transition between outgoing and incoming risk control resources.

⇨ Most companies profess that 'safety is their top priority,' but most safety initiatives are inadequately funded and lack leadership support when safety needs conflict with core business objectives.

- Active Agenda is 'free" to download, use, distribute, and modify. The need for budget approval for Active Agenda will be limited to hosting costs (free if hosted internally) and deployment services. Larger companies have employees that fully recognize Active Agenda's many modules and will 'hit the ground running' once they navigate around their obstacles to innovation. Smaller companies have clerical staff that manually perform many of Active Agenda's tasks on a daily basis and will be grateful to automate such a comprehensive business requirement.

- Active Agenda is an operational performance product, an environmental protection product, a labor relations product, an employee safety product - a business risk control product. By focusing on business tasks rather than functional titles, and integrating 'safety' in core business objectives, Active Agenda can elevate the performance of safety initiatives with more support and less resistance.

A FINAL THOUGHT TO PONDER...

  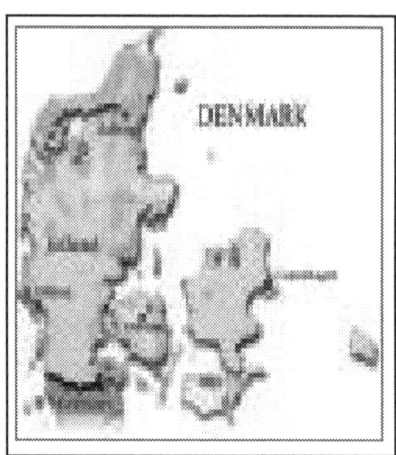

"KANGAROOS EATING ORANGES IN DENMARK"

 **GOOD LUCK IN YOUR QUEST FOR SAFETY EXCELLENCE!**

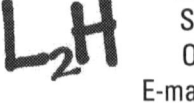 SPEAKING OF SAFETY, INC.
Online at www.L2Hsos.com
E-mail: LLHSOS@dreamscape.com
Tel: 315.383.3801

"ROC YOUR ORGANIZATION!"

Available in:

Soft cover Book,135 pages; Compact Disc (PDF format); and Action Card Deck.

Contains "Insights on Organizational Excellence," (inspiring quotes on Change, Leadership, and Human Performance, from educators, national leaders, business gurus, and progressive safety professionals, and fifty-two (52) ROC's, (Radical Organizational Change) initiatives, ideas, and action mandates to engage an organization's leadership in excellence thinking, and move it toward safety excellence results. Includes the "Test of Excellence" mindset quiz, the "World-Class Safety Strategy" model, the "Safety Excellence Continuum" diagram, and a "Safety Excellence Attributes" self-assessment.

PRICE LIST:

1-5 copies   ---------- Book or Card Deck @ $19.95 US each

6 -10 copies   ---------Book or Card Deck @ $16.95 US each

Over 10 copies --------Book or Card Deck @ $14.95 US each

Compact Disc--------- Book in PDF.format @ $14.95 US each

TO ORDER:

"ROC Your Organization" - Book, CD, or Card Deck:

Complete the following order form and forward to:

Larry L. Hansen
L2H Speaking of Safety, Inc.
Post Office Box 532
Baldwinsville, New York 13027

Note: Payment must accompany orders-- make checks payable to: Larry L. Hansen

NAME _____

TITLE _____

ORGANIZATION_____

STREET ADDRESS OR PO BOX _____

CITY AND STATE_____ZIP_____ _____

COUNTRY_____

NUMBER OF: BOOKS (____); COMPACT DISCS (____); CARD DECKS (____) REQUESTED.

Please add $2.00 for each product ordered (shipping and handling).to the total amount of your order.

Total Enclosed: $_____

Lightning Source UK Ltd.
Milton Keynes UK
UKHW05f0001160418
321101UK00001BA/8/P

9 781933 817118